Management and Industrial Engineering

Series Editor

J. Paulo Davim, Department of Mechanical Engineering, University of Aveiro, Aveiro, Portugal

This series fosters information exchange and discussion on management and industrial engineering and related aspects, namely global management, organizational development and change, strategic management, lean production, performance management, production management, quality engineering, maintenance management, productivity improvement, materials management, human resource management, workforce behavior, innovation and change, technological and organizational flexibility, self-directed work teams, knowledge management, organizational learning, learning organizations, entrepreneurship, sustainable management, etc. The series provides discussion and the exchange of information on principles, strategies, models, techniques, methodologies and applications of management and industrial engineering in the field of the different types of organizational activities. It aims to communicate the latest developments and thinking in what concerns the latest research activity relating to new organizational challenges and changes world-wide. Contributions to this book series are welcome on all subjects related with management and industrial engineering. To submit a proposal or request further information, please contact Professor J. Paulo Davim, Book Series Editor, pdavim@ua.pt

More information about this series at http://www.springer.com/series/11690

Carolina Machado · J. Paulo Davim
Editors

Circular Economy and Engineering

A New Ecologically Efficient Model

Springer

Editors
Carolina Machado
Department of Management
School of Economics and Management
University of Minho, Campus Gualtar
Braga, Portugal

J. Paulo Davim
Department of Mechanical Engineering
University of Aveiro, Campus Santiago
Aveiro, Portugal

ISSN 2365-0532 ISSN 2365-0540 (electronic)
Management and Industrial Engineering
ISBN 978-3-030-43043-6 ISBN 978-3-030-43044-3 (eBook)
https://doi.org/10.1007/978-3-030-43044-3

This Springer imprint is published by the registered company Springer Nature Switzerland AG
The registered company address is: Gewerbestrasse 11, 6330 Cham, Switzerland

Preface

More and more the world faces deep challenges that lead to responsible and sustained use of the resources at our disposal. Replacing the concept of the end of life of the linear economy with new circular flows of reuse, restoration and renewal, in an integrated process, the circular economy is considered as a key element to promote the decoupling between the economic growth and the increase in the consumption of resources, a relationship until now considered as inevitable. In other words, understood as a strategic concept based on the reduction, reuse, recovery and recycling of materials and energy, the circular economy promotes a reorganized economic model through the coordination of production and consumption systems in closed circuits. This way it is easy to see that there are many challenges that organizations are facing. Challenges to the different levels, in so far as a dynamic process, it requires technical and economic compatibility (at the level of capacities and productive activities) but also social and institutional framework (namely at the level of incentives and values). Acting broadly, the circular economy aims the development of new products and services economically viable and ecologically efficient. It is based on minimizing resource extraction, maximizing reuse, increasing efficiency and developing new business models.

Conscious of this reality, this book, entitled *Circular Economy and Engineering: A New Ecologically Efficient Model*, looks to provide a support to academics and researchers, as well as those that operating in the management and engineering fields need to deal with policies and strategies that allow to move towards a more sustainable paradigm, a greener economy that guarantees economic development and the improvement of living and working conditions. Drawing on the latest developments, ideas, research and best practice, this book intends to examine the new advances in the subjects of the circular economy, resultant from the last changes that are taking place and how they affect the management as well as the commitment and motivation of these organizations' workers.

The mission of this book is to provide a channel of communication to disseminate the knowledge of new advances in the circular economy as a new ecologically efficient model and consequently the way how to manage competitive organizations, between academics/researchers, managers and engineers.

Organized in five chapters, *Circular Economy and Engineering: A New Ecologically Efficient Model* looks to discuss in chapter one "The Activities of Recyclable Material Pickers in Brazil and Building Circular Economy Strategies", while the second chapter covers "Effective Economic Decision-Making Methods in Environmental and Sustainability Project Environments and Project Life Cycle". The third chapter focuses the "Development of Policies and Practices of Social Responsibility in Portuguese Companies: Implications of the SA8000 Standard"; the fourth chapter speaks about the "Nexus between Sustainable Development and CSR—An Empirical Study on Indian Nationalized Banks"; and finally the fifth chapter looks to discuss "How to Look at Organizations and Human Resource Management in the Economy of the Future?".

Able to be used by academics, researchers, human resources managers, managers, engineers and other professionals in related matters with circular economy development, this book looks to:

- Share knowledge and insights into circular economy on an international and transnational scale.
- Keep at the forefront of innovative theories, models, processes and strategies, as well as the most recent research activities relating to the circular economy.
- Advance our understanding of circular economy key issues.

The editors acknowledge their gratitude to Springer for this opportunity and for their professional support. Finally, we would like to thank all chapter authors for their interest and availability to work on this project.

Braga, Portugal — Carolina Machado
Aveiro, Portugal — J. Paulo Davim

Contents

Editors and Contributors

About the Editors

Carolina Machado received her Ph.D. degree in Management Sciences (Organizational and Politics Management area/Human Resources Management) from the University of Minho in 1999, Master degree in Management (Strategic Human Resource Management) from the Technical University of Lisbon in 1994 and Degree in Business Administration from the University of Minho in 1989. Teaching in the human resources management subjects since 1989 at University of Minho, she is, since 2004, an associated professor, with experience and research interest areas in the field of human resource management, international human resource management, human resource management in SMEs, training and development, emotional intelligence, management change, knowledge management and management/HRM in the digital age. She is the head of the Department of Management and head of the Human Resources Management Work Group at University of Minho, as well as the chief editor of the International Journal of Applied Management Sciences and Engineering (IJAMSE), guest editor of journals, books editor and book series editor, as well as reviewer in different international prestigious journals. In addition, she has also published both as editor/co-editor and as author/co-author several books, chapters and articles in journals and conferences.
e-mail: carolina@eeg.uminho.pt

J. Paulo Davim received his Ph.D. degree in Mechanical Engineering in 1997, M.Sc. degree in Mechanical Engineering (materials and manufacturing processes) in 1991, Mechanical Engineering degree (5 years) in 1986, from the University of Porto (FEUP), the Aggregate title (Full Habilitation) from the University of Coimbra in 2005 and the D.Sc. (Higher Doctorate) from London Metropolitan University in 2013. He is a senior chartered engineer by the Portuguese Institution of Engineers with an MBA and Specialist titles in Engineering and Industrial Management as well as in Metrology. He is also Eur Ing by FEANI-Brussels and

Fellow (FIET) of the IET-London. Currently, he is a professor at the Department of Mechanical Engineering of the University of Aveiro, Portugal. He is also distinguished as Honorary Professor in several universities/colleges. He has more than 30 years of teaching and research experience in Manufacturing, Materials, Mechanical and Industrial Engineering, with special emphasis on machining and tribology. He has also interest in management, engineering education and higher education for sustainability. He has guided large numbers of postdoc, Ph.D. and master's students as well as has coordinated and participated in several financed research projects. He has received several scientific awards. He has worked as an evaluator of projects for ERC-European Research Council and other international research agencies as well as an examiner of Ph.D. thesis for many universities in different countries. He is the editor in chief of several international journals, guest editor of journals, books editor, book series editor and scientific advisory for many international journals and conferences. Presently, he is an editorial board member of 30 international journals and acts as a reviewer for more than 100 prestigious Web of Science journals. In addition, he has also published as an editor (and co-editor) more than 125 books and as an author (and co-author) more than ten books, 100 chapters and 400 articles in journals and conferences (more than 250 articles in journals indexed in Web of Science core collection/h-index 55+/9500+ citations, SCOPUS/h-index 60+/11500+ citations, Google Scholar/h-index 76+/19000+).
e-mail: pdavim@ua.pt

Contributors

Sudin Bag Department of Business Administration, Vidyasagar University, Midnapore, India

Ana Costa School of Economics and Management, University of Minho, Braga, Portugal

Dimas Estevam University of Extremo Sul Catarinense (UNESC), Criciúma, Brazil

Diana Fernandes School of Economics and Management, University of Minho, Braga, Portugal

Brian J. Galli Hofstra University, Hempstead, NY, USA

Delfina Gomes School of Economics and Management, University of Minho, Braga, Portugal

Carolina Feliciana Machado School of Economics and Management, University of Minho, Braga, Portugal

Atanu Manna Research Scholar, Centre for Environmental Studies, Vidyasagar University, Midnapore, India

Carla Montefusco Departament of Social Service, Federal University of Rio Grande do Norte (UFRGN), Natal, Brazil

Nilanjan Ray School of Management, Adamas University, Kolkata, India

João Leite Ribeiro School of Economics and Management, University of Minho, Braga, Portugal

Maria João Santos School of Economics and Management, University of Lisbon (ISEG/UL), Lisbon, Portugal

The Activities of Recyclable Material Pickers in Brazil and Building Circular Economy Strategies

Carla Montefusco, Maria João Santos, and Dimas Estevam

Abstract This paper adopts the circular economy perspective to analyse the activities undertaken by recyclable waste pickers in Brazil and correspondingly focusing on their working and living conditions. Set in the framework of the Brazilian public policies enacted for the treatment of solid wastes, the findings demonstrate how, despite the recognition of the central role of the pickers as key environmental actors in the national context, their working conditions still remain hazardous and unsanitary, generating scant financial returns and with the social stigma still prevailing. The institutional recognition of the recycling picker profession as key to the management of solid wastes reflects a significant but still insufficient advance towards endowing these professionals with dignified working and living conditions. This also identifies the relevance and importance of the social and associative movements of these workers towards constructing solutions able to value their activities and bring about their social integration. Analysis of the particular case of the recyclable material pickers also conveys the importance of incorporating the circular economy into the socio-economic perspective to the extent that this contributes towards the subsistence of many thousands of people who deserve the guarantee of a minimum standard of living.

Keywords Circular economy · Waste pickers · Public policy · Decent work

C. Montefusco
Departament of Social Service, Federal University of Rio Grande do Norte (UFRGN), Natal, Brazil

M. J. Santos (✉)
School of Economics and Management, University of Lisbon (ISEG/UL), Lisbon, Portugal
e-mail: mjsantos@iseg.ulisboa.pt

D. Estevam
University of Extremo Sul Catarinense (UNESC), Criciúma, Brazil

C. Machado and J. P. Davim (eds.), *Circular Economy and Engineering*, Management and Industrial Engineering,
https://doi.org/10.1007/978-3-030-43044-3_1

1 Introduction

Within the search for alternative solutions able to bring about more sustainable economies, the circular economy (CE) concept emerged. The CE strives not only for the rational usage of resources but also to minimise the need for their generation and the reutilisation of those already in effect. Within this framework, this replaces the linear economic end-of-life concept with new circular flows through reutilisation, restoration and renovation in integrated and systemic processes. The CE, in advocating the decoupling of economic growth from increasing resource consumption, a relationship otherwise perceived as inexorable, redefines the functional flows of chains of production and calls into question the hegemonic model characterised by wastage and disposal.

In order to implement any CE, one decisive facet involves the means of managing waste (discarded materials). Within this framework, recycling plays a structural role in terms of reconverting such wastes into new materials or products with the potential for reutilisation. Given the relevance of recycling to this process, there are currently major demands to boost the rate of waste recovery, especially in cities and the major urban conurbations.

This factor also represents a core challenge to contemporary public policies. In Brazil, after almost a decade since the enactment of the National Solid Wastes Policy (law no 12.305/2010), there still remain major challenges to its effective implementation. This law provides incentives for reducing the generation of solid wastes, selective collection procedures, reverse logistics and the intensification of environmentally appropriate means of waste disposal. However, pickers and their organisations still remain the key actors in this chain given they account for almost 90% of all materials recycled in the country (IPEA, 2013). Hence, there remains a long way to go for the full implementation of the solid waste policy both in terms of selective collection and sorting materials and in the processing and resale of reusable and recyclable waste materials.

The work of pickers takes place in precarious conditions according to the Ministry of the Environment.[1] They may act individually, in an autonomous approach whether on the streets or on garbage dumps, or in collective organisations (cooperatives and associations).[2] Despite such precariousness, the actions of recyclable material pickers have contributed to generating work and income and provide a core input into the solid waste management process in Brazil and fostering sustainable means for handling, reusing and recycling waste materials.

Within this framework, there is a need to question the extent to which greater recognition of the role played by recyclable material pickers in the dynamics of waste reutilisation and recycling networks might result in potential contributions that value their activities and/or reduce their marginal and socially excluded position. Would this

[1] Available at: http://www.mma.gov.br/cidades-sustentaveis/residuos-solidos/catadores-de-materiais-reciclaveis. Accessed on 27 September 2019.

[2] Decree-Law no. 7.217, 21 June 2010, considers cooperatives and associations to be waste processing public service providers.

provide recognition to actors undertaking activities poorly valued by society? Would greater professionalisation of their activities guarantee greater social recognition, decent remuneration and their emergence from the group of the poorest and most excluded? Considering the social and environmental importance of pickers within the current CE paradigm involves perceiving recycling activities not only as technical operations for resource management but also as a relevant socio-economic factor able to contribute to the livelihoods of thousands of people.

In order to better grasp the social facet associated with the recovery of wastes in cities, attention needs paying to the central role of recycling waste in the lives of thousands of poor persons. The dependence of the lesser advantaged on discarded resources extends to a significant extend. Such an activity emerges as a strategy for survival and the generation of income despite the heavy social stigma associated with such activities and in addition to the precarious living and working conditions.

This chapter analyses the activities of recyclable material pickers in Brazil and their contributions to building CE strategies. Focusing analysis on the role of waste pickers as key environmental actors in the Brazilian recycling chain, this then reflects on the extent of such labour activities throughout Brazil before approaching the political and community initiatives that seek to organise the activities of recyclable material pickers and improve their living and working conditions.

Hence, the chapter seeks to contribute to deepening the understanding of the social dimensions associated with the recovery and recycling of urban solid wastes. This portrays the conditions, even while still afflicted by socio-historical adversities, under which recyclable material pickers emerge as important actors in bringing about more sustainable levels of development.

The chapter is structured as follows: following this introductory section, we move on to define the public policy framework in effect in the country and the prevailing economic, social and political context. Subsequently, in the third and fourth sections, we position the work of pickers in the selective collection and recycling systems as well as their different means of organisation and the working conditions that permeate the daily activities of recyclable material pickers. Section five depicts the advances and the challenges to these organisational modes of picker activities based upon the CE perspective in Brazil, with some progress and a lot of outstanding challenges. Finally, the chapter sets out some closing considerations.

2 Recyclable Material Picker Activities in Brazil Measured According to the Logic of Consumption and Disposal

The data on the waste management panorama in Latin America and the Caribbean (UN, 2018) demonstrate that each inhabitant in the region generates around one kilo of waste daily. Thus, across the Latin American and Caribbean region, around 541,000 tons per day are produced with this level of waste production forecast to rise by at least 25% by 2050. In Brazil, 2017 saw the collection of around 71.6 million tons of solid urban waste (SUW), registering a coverage collection rate of 91.2% of the total waste collected. However, the appropriate final disposal of SUW was registered for 59.1% of the total amount. In Brazil, there is a lack of such waste disposal infrastructures, with garbage dumps, present throughout every national region, receiving over 80,000 tons/day of waste (ABRELPE, 2017).

However, according to the Recycling Annual Report (2017–2018), despite difficulties, the selective collection has achieved improvements. The number of municipalities running selective collection systems has more than tripled over a one-decade period, rising from 405 municipalities (7% of the total) in 2008 to 1227 (22% of the total) in 2018 (ANCAT & PRAGMA, 2019).

However, there are also difficulties in providing any precise diagnosis of the selective waste collection situation in Brazil. This reflects in how the Annual Recycling Report (2017–2018) states that of the 5570 Brazilian municipalities, 1227 municipalities operate a structured selective collection organisation while statistics from the National Solid Waste Policy and, simultaneously, from the Profile of Brazilian Municipalities (IBGE, 2017) refers to how 54% of municipalities have an Integrated Solid Waste Plan under implementation. That many urban cleaning services, the collection of solid waste, sweeping and managing processing units which have been outsourced to private companies (accounting for around 70% of total expenditure on urban solid waste management) represents, according to the IPEA (2011), a major obstacle to implementing shared solid waste selective collection and management programs.

Following the publication of the PNRS—the National Solid Waste Policy, no. 12.305/10, there is a formal recognition of the need to implement management mechanisms for the socio-environmental problems caused by inappropriate means of handling solid wastes. The PNRS duly assumes that there are shared responsibilities between manufacturers, citizens and governments for bringing about actions that enable reverse logistics. Precisely such a context has strengthened the role of recyclable material pickers as actors in an activity relatively susceptible to generating important economic and environmental gains—selective collection.

Currently, the most common selective collection models existing in Brazil are door-to-door collection and voluntary recycling deposit bin collection. Door-to-door collection services may be provided by public or private garbage collection services or by associations or cooperative of recyclable material pickers and commonly via vehicles, usually garbage trucks, that follow specified routes on scheduled days in order to pick up waste materials already separated by the population. In turn, the voluntary recycling points are placed in locations situated close to groups of residences for the disposal of household waste materials.

This sets the context for the recyclable material picker segment. Despite the lack of consistent information on the number of people working informally in waste collection, there is evidence that the number of persons involved has been rising. Many of these participants carry out their work in the streets, in favelas or on garbage dumps and are generally individuals from poor, unemployed and socially disadvantaged backgrounds. Thus, their activities take place in precarious and socially undervalued conditions with pickers gathering discarded waste and transforming such materials into their own means of subsistence.

In this context, pickers find themselves placed in an unusual position: while marginalised and generally socially excluded, they perform a key social role as they are responsible for returning products into their life cycles. In this sense, on the one hand, pickers suffer from prejudice due to the fact they work with rubbish and make up one of the poorest sections of the population that lacks access to a series of rights and conditions that only higher levels of income are able to obtain. On the other hand, they undertake a key public utility service essential to obtaining the recycling objectives already defined by the public entities responsible (Magalhães, 2012).

This thus conveys the paradox in the relevance of picker activities within the scope of establishing public policies better aligned with the core assumptions of sustainable development and, in contrast, the lack of value and social stigma still strongly interlinked with such a profession. Simultaneously, holding a central role in the chain of recycling and reutilisation of materials, pickers find themselves in a socially marginalised position facing both difficult working condition and the prejudice and social invisibility that still looms over the activities of these environmental actors on the urban scene.

3 Recyclable Material Pickers: Living and Working Conditions in the Brazilian Reality

In research systematised by IPEA, based on data from IBGE[3]—the Brazilian Institute of Statistics and Geography, there are approximately 388,000 recyclable material pickers across the country. However, the MNCR—the Movement of Recyclable Material Pickers[4] believes there are between 800,000 and one million active pickers. In addition, Dagnino and Johansen (2017) explain that many pickers undertake their collections, classification and sale of materials on a sporadic basis, sometimes to complement the monthly income from another occupation. Hence, it is possible that many such pickers fall beyond the scope of the classification applied by IBGE given that such activities do not fall within the framework of their main working activity.

Furthermore, the actual total number undergoes variations in keeping with the prevailing level of national economic prosperity. In periods of higher levels of formal employment, pickers commonly abandon this activity after having found better

[3]In the 2010 Census, IBGE applied the Classification of Occupations for Household Research (COD), which derives from the CBO—the Brazilian Classification of Occupations collated by the Ministry of Employment (Brazil, 2010) and entitled CBO Domiciliar in Portuguese. This classification contains the following titles and codes of occupation for pickers: main subgroup "96—Rubbish collectors and other basic occupations", which contains the subgroup "961—Rubbish collectors", and, within this, the groups "9511—Rubbish and recyclable material picker", "9612—Classifiers of wastes" and "9613—Sweepers and similar".

[4]Available at http://www.mncr.org.br/sobre-o-mncr/duvidas-frequentes/quantos-catadores-existem-em-atividade-no-Brazil Accessed on 27 September 2019.

working opportunities in other functions. During downturns in the economy, the contrary takes place. Many return to the sector alongside new pickers taking up the activity as a survival option.

In characterising the socio-demographic profile, we find that the overwhelming majority of pickers are male, black (14.6%), mixed-raced (*pardos*) (51.5%), with an average age of 39 and at best hold minimum levels of schooling with a 20% illiteracy rate. The average wage of recycled material pickers in Brazil stands at about the minimum wage (currently, approximately 240 dollars) (Recycling Annual Report, 2017–2018; Dagnino & Johansen, 2017). This general picture serves to convey how undertaking this activity is closely bound up with patterns of self-help and survival. The recourse to waste by the most disadvantaged, such as collecting garbage, fundamentally constitutes a strategy for survival capable of securing the most basic needs of subsistence.

The position of economic vulnerability they face, however, receives no solution from the social security system prevailing in Brazil, itself fragile and limited in scope. As regards receipt of a retirement or pension benefit from the official system (federal, state or municipal), the official data report that out of a total of 398,348 pickers, only 13,858 responded positively, a coverage rate of just 5% (Dagnino & Johansen, 2017).

The social security system is contributory and thus only those who have made payments for the pre-established periods gain any access to working benefits such as maternity pay, remunerated sick leave and retirement pensions. As informal employment still remains common across the waste collection sector and they have low levels of social security payments. Simultaneously, picker incomes, while insufficient, very often do not fall into the category enabling receipt of benefits under Brazilian social protection systems.

It is important to highlight how pickers do not represent any monolithic block as there are those who have engaged in such activities since their childhood while others got involved due to the circumstantial loss of employment and find waste picking as an alternative means of survival. As regards the locations of work, there are those who act along specific routes, going through both residential and retail/business areas with other pickers focusing their activities on the garbage dumps and landfills (IPEA, 2013).

Indeed, in relation to the place of work, 20% of pickers work in their own homes, furthermore reflecting in their non-participation in working associations or cooperatives, and 74% work only in the municipality they reside in but not at home. According to the latest demographic census carried out by IBGE (2010), as regards the largest national regions, the Sudeste region concentrates the greatest number of pickers, accounting for around 42% of the national workforce engaged in this activity, followed by the Nordeste region on 30%. We may observe how, in this period, Brazil had 461 pickers for every 100,000 persons in employment with this ratio rising to 572 in the Nordeste region (Dagnino & Johansen, 2017).

We may highlight that the picker profession generates an insufficient level of income and generic analysis of the conditions prevailing in cooperatives (with a few exceptions) and observes a precarious working scenario that only extrapolates the informality of working relations. The cooperatives still need support to improve their working conditions, for example as regards individual protective equipment, such as gloves and boots, as well as machinery and equipment for developing the activities undertaken, such as presses, scales and means of transport.

Furthermore, they also experience social exclusion and marginalisation, with their labour undervalued due to the dirty and insalubrious working conditions. However, this segregation does not stem from their body odours or the insects they encounter along with the waste materials. Instead, it fundamentally derives from the social marginalism traditionally associated with people handling waste and the very conditions of poverty and social vulnerability that they face. Simultaneously, the garbage collection sector is not rarely interlinked with criminality with waste picking sometimes perceived as "overwhelmed by crime" not only by the authorities but also by the public in general. The dominant perception prevailing in society deepens this conception of self-depreciation. Studies on the self-perceptions of pickers regarding their status report that they rank themselves in the lowest levels of the social hierarchy and generically register dissatisfaction with their occupation (Cherfem, 2015; Costa & Pato, 2016; Sant'ana & Matello, 2016).

As regards their economic position, this correspondingly again highlights the predominantly informal character of these activities. Even those who engage in regular waste collection activities, including the employees of many small waste collection companies, are in the majority outsourced by industries in the formal sector and thus subject to uncertainties and difficulties due to precarious contractual conditions. The working conditions of pickers are dependent on the prevailing recycling chain of Brazil, which both is complex and displays its own specific characteristics that not only vary by region but also from municipality to municipality.

Furthermore, despite pickers and their organisations being the main actors in this chain, given that they are responsible for handling almost 90% of the material recycled in Brazil (IPEA, 2013), they also form the weakest link in the chain, almost always dependent on the actions of third parties and the industries that determine the prices, the volumes and the conditions for the materials acquired (IPEA, 2013). In this context, integrating into cooperatives/associations represents an essential step in the professionalisation of pickers and bringing about more effective organisation of their work.

4 Organising the National Movement of Recyclable Material Pickers in Brazil

Recyclable material pickers have been present in the urban Brazilian panorama ever since the nineteenth century and reflecting how the sector accompanied practically the entire process of national urbanisation. However, the existence of persons living from discarded and collected materials has been more strongly perceived in Brazil since industrialisation. In the early twentieth century, with the expansion of the national printing industry, there are records of collectors of paper for recycling as well as the trade in scrap, especially bottles and metals, in São Paulo. Those collecting such materials became popularly known as "*garrafeiros* (bottlers)" who, over the course of time, gave way to the pickers (Pinhel, 2013).

Given their precarious living conditions and subsistence work, ever since the 1960s various different initiatives have taken place across the respective Brazilian states (many of which have received support from the Catholic Church, non-governmental organisations and universities) to provide support and assistance to groups of pickers and others living and/or working in the street.

However, it was not until this century, with the founding of the MNCR—the National Movement of Recyclable Materials (2001) and the MNPR—the National Movement of the Street Population (2004) that the political organisation of these social groups took place and effectively placing them on the map of national public policies. The first record of a cooperative formed by pickers in Brazil is Coopamare—the Cooperative of Pickers of Paper, Wastes and Reutilisable Materials founded in 1989 in the municipality of São Paulo/SP (Sant'ana & Maetello, 2016; MNCR, 2019).

Following the first National Congress of Recyclable Material Pickers, which took place in 2001 in conjunction with the first National Street Population March, a document entitled "Charter of Brasília" was published and registered the sending of a draft law for regulating the occupational sector as well as the proposal for guaranteeing that the recycling process would undergo development throughout the country with the priority attributed to launching social recyclable material companies.

As a result of the aforementioned processes, 2002 saw the activities of collecting and selecting recyclable material gain recognition in the CBO—the Brazilian Classification of Occupations. Irrespective, such recognition did not bring about any change either to the social stigma of workers dealing with society's waste or even to their informal and insalubrious working terms and conditions.[5]

Based on PNRS, no. 12.305/10, there is a legal framework for the proposed integrated management of solid wastes in Brazil as well as the social inclusion and economic emancipation of pickers in the recycling chain. This reflects the priority the PNRS attributes to sustainable development as a core principle and with the key objective of integrating reusable and recyclable material pickers into actions involving shared responsibilities for the life cycles of products (Art7; XII). Therefore, the PNRS provides for fostering and supporting recyclable material picker cooperatives, deemed fundamental actors to the solid waste management process.

Some of these initiatives constitute part of the response to the precarious working conditions and the exclusion of pickers. However, as the IPC-IG—the International Centre of Inclusive Growth Policies states, there is still no definitive clarity as to whether these initiatives shall bring about the greater empowerment of pickers or if they shall remain excluded across diverse social fields (IPC-IG, 2014).

[5]Activity ranked with the maximum level of insalubriousness by Regulatory Norm no. 15, of the Ministry of Work and Employment (MTE).

5 Recyclable Material Pickers and Their Role in Building Circular Economy Strategies in Brazil: Between Progress and Challenges

Despite pickers being identified as a priority for public policies within the scope of waste management programs, such legislation, however, has to undergo effective implementation (Dagnino & Johansen, 2017). The progress in the regulatory mechanisms has been insufficient to bring about the targeted changes and, not unfrequently, business interests end up prevailing over the public interest. This reflects in medium- and large-scale cities where Cherfem (2015) identifies how they prioritise the contracting of private companies to provide collection services to the detriment of both pickers and their respective associations or instead awarding incentives for waste incineration practices that cause severe consequences for the surrounding environment.

The option to take into consideration the interests of large companies and corporations, benefitting them with long-term contracts for waste collection services, explicitly ignores the principles set out in the PNRS. This plan specifically attributes priority to pickers given the legislative objective of fostering social inclusion and citizenship through means of the recycling chain. Hence, achieving their rights in practice still remains one of the greatest challenges faced by recyclable material pickers in Brazil.

These challenges are not limited to the specific scope of the pickers but rather feature at the centre of Brazilian society itself; still beholden to prejudices around poverty. Therefore, there remains the challenge of ending with the stigma that maintains pickers are socially less important but are rather workers who daily deal with the “rubbish” society produces. Dealing with that discarded means working with that left over, with that no longer socially required. Furthermore, within this same perspective, despite advances in the legal framework, as contained in the PNRS, much remains to be done over its effective implementation. The maintenance of the precarious working and living conditions of Brazilian pickers and the stigma they encounter fully demonstrate this.

This reflection may be extended when placing the emphasis on how the EC might generate new opportunities for employment and income, innovation and the generation of value in Brazil. The vast natural capital of the country may provide for unique EC opportunities, susceptible of simultaneously generating economic, social and natural value (Ellen Macarthur Foundation, 2017). However, to this end, there is the need for a support network involving the state, companies, civil society and the pickers. Hence, this furthermore also duly identifies the essential role played by public policies in nurturing environments favourable to the development of EC networks characterised by citizen participation.

Such a perspective fundamentally requires grasping the life context and social relevance of pickers and thus taking into consideration that this labour activity, in the Brazilian reality, is characterised by the lack of working opportunities. This also needs to highlight, in the processes developed by pickers, the existence of survival strategies that reach beyond the need, and still at an unsatisfactory level, for income. However, to the extent that precarious working conditions continue to prevail within the scope of exercising these activities, the greater the difficulty is in achieving better conditions for the profession.

Furthermore, the need for a broad understanding of the social role of recyclable material pickers remains challenging and even to those who are directly engaged in the activity. The culture of participation and solidarity among members of cooperatives and associations of pickers still requires learning and maturing. Indeed, their participant members are all involved in informal daily working contexts, with low levels of schooling and dealing with a precarious ambience. These difficulties lead pickers to seek out immediate solutions for the resolution of their household needs and, consequently, lacking the time necessary to consolidate cooperative undertakings. Hence, taking the social position of pickers into account is fundamental alongside ensuring an appropriate level of earnings for the families involved (IPEA, 2013).

Despite the challenges referred to, there are signs of change in the legal and regulatory mechanisms as well as in the expression of citizenship in Brazil. The recognition of pickers as a profession and as central actors in the management of solid wastes represents a significant advance even if not sufficient to ensuring these professionals decent working conditions and income dignity. This furthermore highlights the importance of social and associative movements in the construction of solutions that strengthen the human dignity of their members and value the profession. Hence, in the midst of picker activities, there are the first shoots, even while still very green, of new and potential paths for building community-based economic strategies focused on the circularity of resources.

6 Final Considerations

In the Brazilian case, following the enactment of the PNRS in 2010, there has been a progress in defining and implementing solid waste management plans and strengthening the chain of value of recycling activities. Nevertheless, we may also point to the shortage and weakness of public entities and organisms in Brazil to better and more effectively control compliance with the aforementioned legal stipulations.

We may also infer that the facet of inequality that makes up the Brazilian socio-historical background still remains very present and broadly reproduced as in the particular case of recyclable material pickers. Despite their relevance in this process, responsible for handling almost 90% of recycled materials in Brazil, contributing towards generating significant (although not yet calculated) economies in resources and energy, their activities still continue to be marginalised and undervalued.

In a contradictory fashion, the key actors in the Brazilian chain of recycling hold the least value in the processes whether in terms of the income they receive or their prevailing level of social recognition. The process of class segregation, a structural facet of Brazilian society, is revisited in the conception that pickers are less important actors as they deal with the discarded waste of consumerism.

The recyclable material picker organisations (associations and cooperatives) have been founded as a means of political resistance to such segregationist values and environmental degradation. However, as the IPC-IG states, there is still no clear perception as to whether these initiatives shall effectively contribute to the greater empowerment of pickers or if their position of exclusion and marginalisation shall remain coupled with subsistence-based strategies.

This indicates how there is a need for future research to further study the social dimensions associated with solid waste management. This assumes the recognition of the importance of these activities in the livelihoods of poor and disadvantaged households but also the fact that in the majority of cases this also interlinks with problems of poverty, social exclusion, poor social status and health risks.

Defining broader reaching and more socially responsible public policies depends on better understanding and knowledge about the social implications associated with waste management and increasing the social awareness about such problems. This equally depends on research into the means and forms of cooperation that shall enable government agencies, community organisations, voluntary groups and the public in general to work together to generate substantial changes in the social components associated with solid waste management.

Therefore, any interventions need to be analysed within the perspective of how they contribute towards aiding pickers to meet their basic needs and improve their economic position while reducing exploitation and discrimination. Broader reaching and socially more responsible waste management policies should therefore incorporate this objective of providing for decent standards of living. Additionally, implementing this would also mean ensuring the definition and implementation of integrated and sustainable CE strategies.

References

ANCAT & FRAGMA. (2019). *Anuário da Reciclagem (2017–2018)*. Accessed on de outubro 10, 2019 from https://ancat.org.br/wp-content/uploads/2019/09/Anua%CC%81rio-da-Reciclagem.pdf.

ABRELPE—Associação Brasileira de Empresas de Limpeza Pública e Resíduos Especiais. (2017). *Panorama dos Resíduos Sólidos no Brasil*. Brasil: ABRELPE.

Cherfem, C. O. (2015). A coleta seletiva e as contradições para a inclusão de catadoras e catadores de materiais recicláveis: Construção de indicadores sociais. *Mercado de Trabalho, 59*(21), 89–98.

da Costa, C. M. & Pato, C. A. (2016). Constituição de Catadores de Material Reciclável: A identidade estigmatizada pela exclusão e a construção da emancipação como forma de transcendência. In Pereira, B. C. J., & Goes, F. L. (Orgs.), *Catadores de Materiais Recicláveis: um encontro nacional*. Rio de Janeiro: IPEA.

de Dagnino, R. S., & Johansen, I. C. (2017). *Os Catadores no Brasil: Características demográficas e socioeconômicas dos coletores de material reciclável, classificadores de resíduos e varredores a partir do censo demográfico de 2010*. Economia Solidária e Políticas Públicas, Mercado de Trabalho, 62. Accessed on June, 27 2018 from http://repositorio.ipea.gov.br/bitstream/11058/7819/1/bmt_62_catadores.pdf.

Ellen Macarthur Foundation. (2017). *Uma Economia Circular no Brasil: Uma abordagem exploratória inicial*. Brasil: CE100.

IBGE—Instituto Brasileiro de Geografia e Estatistica. (2017). *Perfil dos municípios brasileiro*. Coordenação de População e Indicadores Sociais. Rio de Janeiro: IBGE.

IPEA—Instituto de Pesquisa Economia Aplicada. (2011). *Diagnóstico sobre os catadores de resíduos sólidos*. Brasília: IPEA.

IPEA—Instituto de Pesquisa Economia Aplicada. (2013). *Situação Social das Catadoras e dos Catadores de Material Reciclável e Reutilizável*. Brasília: IPEA.

Magalhães, B. (2012). *Liminaridade e exclusão: Os catadores de materiais recicláveis e suas relações com a sociedade brasileira*. Dissertação (Mestrado em Antropologia). Belo Horizonte: Faculdade de Filosofia e Ciências Humanas, Universidade Federal de Minas Gerais.

Ministério do Meio Ambiente. (n.d.). *Coleta Seletiva*. Ministério do Meio Ambiente. Accessed on October 10, 2019 from https://www.mma.gov.br/cidades-sustentaveis/residuos-solidos/catadores-de-materiais-reciclaveis/reciclagem-e-reaproveitamento.

MNCR. (2019). Movimento Nacional dos Catadores de Materiais Recicláveis. Accessed on October 10, 2019 from http://www.mncr.org.br.

Pinhel, J. R. (org.). (2013). *Do Lixo à Cidadania*—Guia de Formação de Cooperativas de Materiais Recicláveis. Accessed on October 02, 2019 from http://revista.oswaldocruz.br/Content/pdf/Edicao_16_SILVA_Monique_N.pdf.

Política Nacional de Resíduos Sólidos. (2010), Lei N° 12.305 de 02 de agosto de 2010—Política Nacional de Resíduos Sólidos (PNRS).

Sant'ana, D. D., & Maetello, D. (2016). Reciclagem e Inclusão Social no Brasil: Balanços e desafios. In Pereira, B. C. J. & Goes, F. L. (Orgs.) *Catadores de materiais recicláveis*: *Um encontro nacional* (pp 21–44). Rio de Janeiro: IPEA.

Effective Economic Decision-Making Methods in Environmental and Sustainability Project Environments and Project Life Cycle

Brian J. Galli

Abstract Effective economic decision-making is a critical aspect in private or public organizations, especially when it comes to environmental and sustainability project initiatives. Executives, project managers, and teams make decisions daily about the exchange, generating new ideas, data review, and the evaluation of substitute approaches and policy implementation. Furthermore, project environments greatly determine the nature of these decisions. This study examines the magnitude to which decision-making influences organizations to achieve their goals through their daily decisions. The study employed a descriptive and qualitative research design. Also, a questionnaire was used for data collection and was designed using a 5-point scale that ranged from "strongly agree" to "strongly disagree." The findings were that organizations with competent teams had ideal decision-making structures that traded along with desired strategic management practices. Moreover, young organizations relied on the services of consultants for advice regarding complex decisions. The nature of project environments equally influences the decision-making process for project managers within their respective organizations. Investing in information analysis was also crucial for effective decision-making in organizations. Equally, organizations took a careful approach and invested in post-decision activities, such as follow-ups and timely monitoring of the implementation of decisions. Such was crucial in instituting performance checks and progress monitoring of projects. Finally, the study revealed that effective decision-making was a result of systemic processes that involved people, systems, and resources within organizations.

Keywords Decision-making · Project environment · Project life cycle

B. J. Galli (✉)
Hofstra University, Hempstead, NY, USA
e-mail: brian.j.galli@hofstra.edu

C. Machado and J. P. Davim (eds.), *Circular Economy and Engineering*,
Management and Industrial Engineering,
https://doi.org/10.1007/978-3-030-43044-3_2

1 Introduction

1.1 Background

As the world changes to being hyper-connected because of globalization, are organizations reinventing themselves to enjoy a competitive edge? As such, organizations, through effective decision-making structures, are ever-striving to adapt to the dynamic environment to survive. The fact is that the contemporary environment is so competitive and characterized by mergers, downscaling, acquisitions, and uncertainty that various project leaders are on toe to establish ideal ways to improve performance and to decrease competitiveness (Besner & Hobbs, 2012; Al-Kadeem, Backar, Eldardiry, & Haddad, 2017; Obi & Agwu, 2017).

The advantages of effective decision-making approaches continue to receive unlimited support, especially during project implementation. With a rise in competition, organizations continue to be under pressure with the ever-increasing pace of change (Besner & Hobbs, 2012; Eskerod & Blichfeldt, 2005; Marcelino-Sádaba, Pérez-Ezcurdia, Lazcano, & Villanueva, 2014; Svejvig & Andersen, 2015). For instance, firms are faced with multiple choices that range from investment decisions to recruitment decisions. Besides, Besner and Hobbs (2012) assert that a firm's efficiency is solely dependent on its prevailing stability and decision flexibility. Such may make a difference during project management, as research suggests that every success or failure results from making a decision or not.

Thus, an organization's efficiency in decision-making methods reflects openly in executing them swiftly, so firms require a working procedure to be in line with the dynamic environment (Grünig & Kuhn, 2009; Zwikael, & Smyrk, 2012; Xue, Baron, & Esteban, 2017; Schwedes, Riedel, & Dziekan, 2017). Besides, decision-making helps organizations to be on track, which determines a firm's success and failure, as noted by Usman Tariq (2013). For example, the moment that Apple made a decision not to license its operating system to other prevailing manufacturers, it lost the opportunity to enjoy a monopoly, such as Microsoft (Sharon, Weck, & Dori, 2013; Papke-Shields & Boyer-Wright, 2017; Usman Tariq, 2013). This study is interested in establishing effective economic decision-making methods in project environments and project life cycle.

1.2 Effective Economic Decision-Making Methods

The term "decision-making" serves to invoke a choice amidst alternative courses of action in a manner that applies to the prevailing situation demand. The capability of a given decision-maker to arrive at the best option demands an organized decision-making approach (Xue et al., 2017; Lee, Lapira, Bagheri, & Kao, 2013; Galli & Hernandez-Lopez, 2018; Cova & Salle, 2005; David, David, & David, 2017). This section outlines some of the most effective structures of effective decision-making approaches that project managers and leaders should follow to help in setting goals for a project, regardless of the prevailing project environment. Such decision-making road maps are meant to cut on the costs and to maximize gains.

Quality and speedy decision-making improve performance for organizations in projects. As such, directors must strive not to make effective decisions singlehandedly (Xue et al., 2017; Andersen, 2014; Brown & Eisenhardt, 1995; Galli, 2018a; Labedz & Gray, 2013). This can be realized in several ways. First, one can nurture the decision-making skills of their working teams that may be comprised of senior management teams. Second, one can support decisions that are in line with set project goals and the ultimate corporate strategy. Finally, one can facilitate the implementation of such decisions.

Below is an outline of nine applicable steps to be emulated by project managers and leaders of contemporary organizations to enhance their daily decision-making (Xue et al., 2017; Hartono, Wijaya, & Arini, 2014; Parker, Parsons, & Isharyanto, 2015; Xue, Baron, & Esteban, 2016):

- Do not make every decision: Only unproven project managers make all decisions, no matter how small. Directors should be making decisions regarding project strategy, hiring and firing, and resource allocation, which have bigger impacts on projects that they lead. There is a need to trust those working under one's leadership to assist in making decisions. Encourage them to make decisions at their levels and only consult when necessary, provided they have the desired expertise and authority to do it.
- Make your people take a position: If senior project managers request a discussion over a decision, compel them to suggest a well-considered position. If you have the right team, then most of them may be smarter than you regarding their fields of expertise. Thus, for effective decision-making, directors will always rely on the expertise and ideas of their senior management teams.
- Act Swiftly: Organizational directors ought to be comfortable with decision-making, especially in times of crisis, without having to wait for information from different departments. Doing so would render the project susceptible to lose on opportunities or to make losses.
- Change bad decisions quickly: Rarely do project managers want to admit their failures, but again, failing to change such decisions is even more dangerous. To preserve credibility and efficacy, project leaders ought to reverse their bad decisions immediately before things become worse. Do not patch up bad decisions, as they will continue affecting project implementation, which will waste available resources that would be significant in other project areas.
- Assign a devil's advocate: Certain decisions, such as regarding major acquisition, are next to impossible to be reversed and to call for carrying a tremendous risk. Thus, there is a need for a well-thought discussion and analysis. A senior manager can play the devil's advocate to test conclusions and to identify any possible weaknesses, which will help in decision-making.
- Communicate the correct information on time: Rumors and distortions begin taking shape the moment that your management team learns of your decisions by

secondhand. Always communicate in detail about the decisions you make directly to your management team. Let some of your talented managers be aware of the bottom-line of the decisions you make for comprehension's sake, their support, and to facilitating buy-in. This will help your teams in future decision-making.

- Support your people, unless they are wrong: If you encourage your management teams to come up with solutions that you support, give them credit if they excel. Also, stand in solidarity with them if things turn out contrary to expectations. Nevertheless, at times, you may need to make difficult choices that contradict the consensus, but it is significant to explain the reasons for overruling everyone.
- Do not overrule decisions by your team often: If your team applies the organization's strategy, vision, and goals while making decisions, chances are higher that they will make the right decisions. Nevertheless, if you are always at odds with your team, a problem could be in the offing. Thus, the frequent overruling of your team's decision is not appropriate, while striking consensus at the same time is not easy. Having to balance between two extremes is the bottom-line to success. As a leader, there are times that you ought to arrive at unilateral decisions and just move on. Only if a decision fails to consider the time aspect as critical would you need to try and build a consensus.
- Conduct an official postmortem: The only way of establishing if a given decision was a right one is through having a formal postmortem. Effective decisions ought to be re-evaluated by a time-to-time review of significant metrics and the ultimate performance. In the absence of postmortem, re-examining issues in decisions or learning lessons from such decisions becomes difficult. Holding official postmortem helps project leaders with enhanced skills regarding decision-making.

Although it is difficult for leaders and project managers to always have perfect decision-making reputations, following the above nine steps may aid in facilitating better decisions that would steer project managers to achieve set project objectives (Xue et al., 2017; Zhang, Bao, Wang, & Skitmore, 2016; Todorović, Petrović, Mihić, Obradović, & Bushuyev, 2015; Winter, Andersen, Elvin, & Levene, 2006; Sutherland, 2004). Having a structured process for decision-making at any level helps project managers and leaders to meet set organizational goals.

1.3 Environment and Sustainability

To comprehend environmental sustainability, it is necessary to first look at "sustainability." Sustainability is being able to persist on a distinct behavior indefinitely. For an efficient definition, three areas must be addressed in environmental sustainability: one, for pollution, two, for renewable resources, and three, for non-renewable resources (Lamming & Hampson, 2000). Regarding pollution, the rates of waste generation from company projects should not, at any time, surpass the assimilative capacity of its surroundings, which is sustainable waste disposal.

Goodland (2005) wrote that when it comes to renewable resources, the harvest rate should never go beyond that of regeneration. For those resources that are non-renewable, their depletion requires the equivalent development of renewable substitutes for similar resources. Thus, the standard elucidation of environmental sustainability is sustainable development and can only be achieved with sustainable economic development (Lamming & Hampson, 2000). However, there is no economic progress that can progress indefinitely.

It must be understood that companies do not participate in sustaining the environment. However, we know the importance of the environment being sustained so that consumers of the manufactured products can live to enjoy the products. Once raw materials are extracted from the same environment, measures must be implemented to ensure the extracted materials are replaced (Goodland, 2005). The replacement can only be possible through sustainability. A fair example is a paper mill. For a paper mill to thrive, there must be plenty of trees. If the paper manufacturing companies just cut trees without planting more, the companies will not last when all the trees are gone. To avoid such a case, there must be effective economic decision-making and project management strategies to ensure more trees are in the fields, and that they are taken care of until they reach maturity to be harvested by paper mills (Lamming & Hampson, 2000).

Some improvement strategies for the paper mills include planting trees that do not take decades to grow and minimizing the number of trees cut in a given area. The companies can also urge clients to give back to the environment by planting more trees in residential areas. Some of the used books, newspapers, envelopes, and magazines are usually thrown away or burned. When the paper mill company urges people to take these products to recycling centers, an important objective to sustain the environment is achieved (Goodland, 2005). This way, the company can offer quality end products to the customers while maintaining the environment.

1.4 Project Environments

Project environments simply refer to organizational, cultural, and social environments under which a given project is run. Such a process entails a project manager's need to identify project stakeholders, as well as their potential to influence the project's outcome (Yun, Choi, Oliveira, Mulva, & Kang, 2016; Nagel, 2015; Medina & Medina, 2015; Hoon Kwak, & Dixon, 2008; Galli, Kaviani, Bottani, & Murino, 2017). As such, the project manager has no choice but to work hand-in-hand with people to realize desired results. This is more so the case in highly technical and complex environments that may involve contemporary projects. Hence, it is prudent for a project manager and for the project team to be comfortable with the project's organizational, cultural, and social surroundings.

Having a better understanding of project environments places the project manager in a better position, as he can influence the project environment positively for better reception of the project change (Burnes, 2014; Arumugam, 2016; Badi & Pryke,

2016; Ahern, Leavy, & Byrne, 2014). For instance, some people will resist change more than some stakeholders within the project environment. Other stakeholders could have interests that could be indirectly associated with the project. If such issues are realized early enough, they should be handled in a proactive manner that manages the corresponding risks of the project.

1.5 Project Lifecycle

The project lifecycle refers to an assortment of project phases that are subdivided and equally assigned for the appropriate control, management, and operation. Project lifecycles are normally different depending on the project nature and characteristics and the implementing organization. As such, project lifecycles are usually meant to connect a distinct project phase from beginning to end. Hence, the interphase transition may be actualized using some aspect of the transfer technique: handoff or practices of corresponding phases.

The following are the prevalent outcomes that are normally defined through project lifecycle:

- The kind of technical work to be undertaken in each phase when every phase must produce an outcome, how they are to be reviewed, verified, validated, who is to be involved in every phase, and their responsibilities are elements of the project. Lastly, there is project inspection, control, and approval.

Therefore, characteristics of the project lifecycle are usually dependent on project nature and features, though they share some of the below attributes:

- Phases are successive. Transitions normally are effected through technical information transfer or via technical component handoff.
- Cost and staffing cycle is normally at the top during intermediate phases and usually low at the beginning and end of projects.
- Certainty level for completion is usually high at the moment that the project is progressing and not when it begins.
- In contrast, the more that there is an increase in the cost of change, the more correcting errors also increase as the project progresses. In such a situation, the influence attributed to a stakeholder decreases quite dramatically when compared to the initial project phase.

1.6 Problem Statement

Firms seek effective and cost-efficient avenues to deliver high-quality services due to the scarce nature of resources, the ever-increasing demands, and the budget constraint

(Detert, 2000; Easton & Rosenzweig, 2012; Galli, 2018b, c). Effective decision-making approaches serve as integral aspects in project environments while ensuring that the implementing organizations well actualize the project lifecycle. With the dynamism of project environments, organizations and project managers are increasingly faced with challenging situations that subject them to pressure. Consequently, Galli (2018c) reveals that, with heightened competition, firms strive to establish effective decision-making structures to implement their strategy, while meeting set project objectives and goals.

As such, firms that have unfitting strategies fail to plan appropriately. In the end, project objectives are never met, and organizations may close up. Thus, effective decision-making leads organizations in finding sufficient strategies and plans for implementation. This is achieved through teamwork and idea-sharing among management teams while following a structured decision-making procedure. Hence, Galli (2018c) asserts that a firm's success in implementing its strategy lies in its planning process. Formal planning that involves effective decision-making enables organizations to have plans that inform their strategies, so decision-making, strategies, plans, and projects are born. Finally, control measures and progress monitoring is done, which steers firms to realize set project goals, such as profit-making (Galli & Kaviani, 2018; Gimenez-Espin, 2013; Loyd, 2016; Milner, 2016; Parast, 2011). Decisions must be well planned with a clear reflection in regard to the project environments and the project lifecycle. Therefore, it is pretty clear that challenges faced by corporate leaders and project managers are attributed to failures in the decision-making and the planning process.

1.7 Research Objective

Literature covers the important role of these variables, their concepts, and models in project management and performance. However, this study seeks to establish the likenesses and differences within the variables, concepts, and models. This study also aims to devise a framework of the best practices to generate a comprehensive and "universal" framework for all projects, operations, and performance elements. Furthermore, this study supplies evidence-based answers to any optimal questions from experts on the variables, concepts, and models, such as how to best use them for successful projects and objectives. Lastly, the findings of this study can act as a platform for future research. This study sought to show that there is no significant correlation between effective economic decision-making methods and project environments and the project cycle.

1.8 Originality

For projects to be successfully undertaken within diverse environments, decisions made by the project managers are crucial. Thus, this study will serve as a reference point, especially for organizations implementing projects and struggling with the decision-making aspect that may affect generating set project goals and priorities.

Besides, with the ever-increasing demand for effective decision-making within project environments, the survey shall detail an effective decision-making plan that organizations with various projects may apply with project lifecycles. Further, the study explores the effect that economics and effective decisions made in multiple project environments can have on the project lifecycle.

Data within this paper are derived from other studies that have tested these hypotheses, and many research perspectives are added to offer novel approaches to current problems. Firstly, there is a design-science-investigate approach to this study. Secondly, this study supports a valuable growth reveal for any reasonable and hypothetical application. Thirdly, there is an assessment model for the variables, their concepts, and models with a focus on the evaluation instruments to respond to the examination question. Also, there is an outline of the development models, while the evaluation instrument is reviewed, as well. The analysis features an outline of consequences from the meetings, and the conclusion features findings or suggestions that coordinate investigative limitations and future research options.

Much literature highlights the importance of these variables, concepts, and models in project management and operations management. Thus, this paper contributes to the profession, while the findings show the benefits and detriments when performance and sustainability are not considered. Lastly, the true-to-life examples show the importance of assessing these theories not only in theory but also in practice.

1.9 Contribution to the IE/EM/PM Profession and Research Fields

This study contributes to industrial engineering (IE) research, as engineers can improve their work process with this study. The information can expedite the work process for an engineer and can help to save on additional resources: time, money, materials, energy, work hours, etc. Practitioners can also find new ideas in the model of this study to help with creating products. Industrialists can also find helpful information for the research field. This study is written with simple vocabulary, and the framework is comprehensive to act as a future reference for any study.

The study will be of significance to the following:

- Project Managers: Through this study, managers heading different projects will benefit from the proposed effective decision-making approaches that will add value to making decisions, which will steer their respective projects to success.

- Research field and academic world: The study will add value to the existing body of knowledge both in the research and academic worlds. The outcome of the research will be used or referred to by other researchers doing studies in the project management field. Equally, the study will recommend areas for future research.

1.10 Organizational and Managerial Contribution and Relevance

An engineering manager is required to make decisions, and this will be even more important in the future for project management and engineering. The vital implications in the findings are concentrated on many organizational levels (i.e., the corporate level, the managerial level, and the project team). The conclusion allows the engineering management practitioner to find information for capitalizing on these variables, concepts, models, and their relationships.

Also, this study's results apply to many business subjects, as they contribute to current literature that features a research gap. An assessment of the relationships between the variables, concepts, and models allows their advantages and disadvantages to be understood. Thus, the variables, concepts, and models can be used more effectively. Another objective of this study is to uncover more perspectives, as practitioners can better utilize these factors and their relationship with their strategies.

1.11 Paper Organization

The sections of this study are as follows: Section two is a high-level literature review of current literature in these research fields. Section three is the research methodology, and section four presents this study's findings. Finally, section five concentrates on the implications for the practitioner, and it provides new approaches for future research, reveals research limitations, and makes general conclusions.

2 Literature Review

The aspect of decision-making is increasing in significance in private life and organizations. Galli (2018c) notes "Decision-making lies at the heart of our personal and professional lives." For instance, some entities cannot arrive at effective decisions and act on decisions to lose ground (Shenhar & Levy, 2007; Hamel, 2006). Thus, organizations that execute critical decisions proficiently outperform their counterparts that more slowly implement brilliant decisions (Rogers & Blenko, 2006; Shenhar & Levy,

2007). Being a science, decision-making has been a raging subject for some time, more so in the discipline of economics and management literature, as is evident in the decision theory (Gänswein, 2011; Von Thiele Schwarz, 2017). Further, in a business context, decisions are made to enhance growth and profitability. For project management, decisions are made right from project conception up to project completion. Thus, at whatever level or field, decision-making is meant to ensure that set objectives are met and that desired outcomes are attained at the end of projects/businesses.

2.1 Theoretical Review

Decisions that are either made by organizations or individuals may be subdivided into eight categories (Shenhar & Levy, 2007; Xiong, Zhao, Yuan, & Luo, 2017). Every category attempts to portray the nature, significance, or every decision's duration. The seven categories of decisions made include major decisions, minor decisions, programmed decisions, non-programmed decisions, group decisions, individual decisions, and strategic decisions. Below are several decision-making models applicable to the study.

2.1.1 Analytical Hierarchy Process (AHP) Model

Several models exist in the literature that provide frameworks and tools that may help to streamline decision-making into a clear and formal process. This model, as advanced by Saaty (2008), helps to arrive at analytical decisions by evaluating various alternatives to follow set criteria. As such, AHP is a pairwise comparison model that makes use of expert judgment to arrive at priority scales, while establishing a mathematical analysis of likely available options (Saaty, 2008). In the course, the decision problem is shown in a hierarchy that is comprised of three levels: the goal that informs a given decision, the criterion for arriving at alternatives, and the basis of such decisions. Accordingly, every alternative is then assigned with a weighted criteria value that shows how it fits into established criteria. Generally, while AHP leads to good results, it is a laborious model to apply (Saaty, 2008). Thus, it may easily fit in making big, one-time decisions, but it is not ideal for daily operation-based decisions. This study aims to establish a decision-making approach that may be generally applied daily to make project-based decisions. Thus, AHP is based on making bigger decisions that only take place once. The model offers a suitable formal method that helps in defining options and evaluating how such goals may be obtained while focusing on the outcome. Therefore, the AHPs decision-making hierarchy serves as a powerful tool for visualizing how fitting the defined options meet set criteria.

2.1.2 Decision Analysis Model

Although it faces similar criticisms for being laborious, the decision analysis, as brought forth by Howard (1988), approaches an issue that requires a decision in a "decision problem" by following three steps: formulate, evaluate, and appraise. The basics are quite applicable, but the multiple processes to be covered require much time. The model's ultimate framework is rational, as it offers an ideal structure for one's thinking. Essentially, the model's final aim is developing a solution that best fits the situation at hand (Howard, 1988; Ulbrich, 2010). Thus, the entire process of the Decision Analysis theory avails tools for analysis' sake and structured thinking.

2.1.3 General Heuristic Decision-Making Model

Grünig and Kuhn (2009) examine the aspect of the entire decision-making process while upholding a practical point of view. In their model, they define a seven-step approach referred to as "the general heuristic decision-making process." Such a process encompasses simple phases from problem identification to a conclusion. They also provide an easy framework that may be applicable in project management when encountering lower costs in the process. The process is equally applicable and detailed to cover significant aspects without necessarily over-emphasizing on smaller issues. Thus, the 7-step general heuristic decision-making process includes:

1. Discovering the decision problem;
2. Analyzing the decision problem;
3. Developing at least two options;
4. Defining the decision criteria;
5. Drawing up possible scenarios where possible;
6. Determining the consequences of the options;
7. Establishing overall consequences and making final decisions.

For this model, the terminology "heuristic" is employed to refer to a "rule of thumb": a strategy that means almost the non-existence of official application and low application costs in the event of a heuristic model (Grünig & Kuhn, 2009; Wideman, 1990). As contrary to the previously discussed two models, the heuristic model is not meant for solving a specific problem, as it applies to general decision-making, regardless of the nature of the issue.

Hence, the model's two qualities make it ideal to be applied to different kinds of decisions and situations within organizations. As much as the model's process is criticized for low-quality enhancing decisions, it is a bit practical and includes all critical aspects. Thus, it is an effective decision-making approach that is worth using for project managers and leaders.

2.2 Using Game Theory to Enhance Decision-Making

Most corporate leaders and project managers continue to utilize the application of game theory to make high-level/risky decisions, more so during highly competitive situations (Buchanan & O'Connell, 2006). This very model has been in existence for over 50 years and has been proven capable of generating an ideal strategic choice in multiple complex situations of decision-making.

2.3 Colleague-Role Approach to Executive Decision-Making Theory

Making executive decisions within organizations refers to the act of making decisions that have consequences for successive organizational activities. Individual members seldom undertake such a decision, as individuals work in project teams or task forces by coordinating their efforts through exchanging ideas for realizing set objectives. Despite the significance of such interactions amidst those working in organizations, such efforts receive little attention. Mostly, reference is directed to the existence of "informal organization" and its significance, though little insight is gained as to how such interactions help the decision-making process (Hammond, Keeney, & Raiffa, 2006).

2.4 History of Effective Decision-Making

The terminology "decision-making" began in the mid-eighteenth century by Chester Bernard, the author of "Functions of the Executive." Bernard was a retiree and a telephone executive who imported the terminology "decision-making" from a public administration dictionary to the corporate world. Hence, the terminology began replacing the narrower descriptions, such as "resource allocation" or "policymaking" (Buchanan & O'Connell, 2006).

As a result, by the mere introduction of the "decision-making" phrase, he altered the way that corporate leaders thought about their deeds, ushered in a new era of action, and created a desire for conduciveness from corporate leaders and project managers. "Decision" basically means the end of deliberating and the beginning of the action.

Bernard, and the later contribution of theorists, such as Henry Mintzberg, James March, and Herbert Simon, formed the basis for studying managerial decision-making in the project management and business world. Moreover, decision-making as a field of study is an intellectual discipline that brings together other fields of psychology, political science, sociology, economics, and mathematics. For instance, philosophers tend to deliberate what people's decisions say about them and their values. Historians dichotomize the choices that leaders make during critical junctures. Also, research regarding risk and organizational behavior emanates from a practical craving to help managers realize enhanced outcomes. Whereas good decisions do not guarantee a good outcome, such rationality eventually pays off. Further, there is a growing complexity with having to manage a risk enhanced understanding of human behavior and technological advances that aid cognitive processes to improve decision-making (Albert, 2006; Steinburner, 1974).

The history of decision-making approaches has never been of utter progress toward attaining perfect rationalism. For years, managers and corporate leaders increasingly continue to comprehend the constraints (both psychological and contextual) on their potential of making optimal choices and better decisions. Other decision-making entities within organizations assert that intricate circumstances, limited time, and insufficient mental computational power tend to reduce decision-makers to a level of "bounded rationality." Others reveal that managers and organizational leaders would only make economically rational decisions upon gathering sufficient and relevant information.

Decision-making's administrative behavior theory, as advanced by Easton and Rosenzweig (2012), dwells on studying the decision-making process within administrative firms. According to Herbert, decision-making refers to the heart of the administration, and administrative vocabulary ought to be derived from logic and the psychology of human choice. He attempts to pronounce administrative firms in a manner that provides the rationale for doing scientific analysis. Moreover, Herbert is against the idea of having an omniscient "economic man" with the ability to make decisions, which may bring the greatest benefit, possible. Instead, Herbert replaces such an ideology with the notion of having an "administrative man" who optimizes, instead of maximizes, his decision effort. According to Herbert, there is not a single sure way of managing or giving the best decision. Rather, he advocates for the view that the decisions made by corporate leaders are just good enough, but are never the best because of subjective human elements that intervene during the decision-making discourse.

As such, Herbert concludes that decisions made by managers in the corporate world are "satisfying" and good enough, as opposed to "maximizing," which is the best decision. Such is supported by the notion of "bounded rationality," which refers to the thinking that the rationality of people is usually limited by the nature of the information at their disposal during decision-making through cognitive limitations. As such, the decision-makers (despite their intelligence levels) are obliged to work under three unavoidable controls: the limited nature of the information at the disposal of decision-makers, the restrained capacity of the human mind in evaluating situations, and the limited nature of time that is available for the decision-making process.

Moreover, Mansfield (1999) surveyed the behavioral perspective regarding the theory of the firm that explains the systemic-anarchic of a firm's decision-making, which is famously referred to as the garbage can model. Although Mansfield's work scope was extensive, he eventually focused on understanding how decisions take place within individuals, organizations, and groups in society.

In Henry Mintzberg's groundbreaking article, "The Nature of Managerial work" (1973), he established what a manager's work entails. For example, managers shoulder a large workload by necessitating interruptions, giving quick responses to each stimulus, seeking the tangible, avoiding the abstract, and ensuring that everything is done abruptly. As such, Mintzberg suggests six characteristics that define managerial work. They equally apply to managerial jobs (from project managers to the executive leaders) in the corporate world. The six attributes include:

1. Manager's job descriptions comprise of a blend of programmed, regular, and un-programmed tasks in the organizations they serve.
2. Managers are both specialist and generalist leaders within their organizations.
3. Managers depend on information from wide sources, but they eventually exhibit a preference for the only one that can be orally transmitted.
4. Managerial work comprises of activities that are defined by brevity, variety, and fragmentation.
5. Management work is more of an art than a science, and it entirely relies on intuitive processes.
6. Day by day, management work is becoming ever complex.

2.5 Characteristics of Information During Decision-Making

In the course of the effective decision-making approach, analysis happens to be identified as a significant aspect. It lays bare the rationale on which a given decision is arrived and defining the nature of available options for consideration (Malhotra, 2005; Campbell, Whitehead, & Finkelstein, 2009). Generally, during analysis, the nature of available information occupies a significant role. Thus, this study delves into establishing some universal attributes of information during decision-making. Therefore, Malhotra (2005) asserts that those in the decision-making position usually expect the available information about what is before them for a decision to be accurate, current, sufficient, available, and equally relevant. As much as all aspects may not be met regarding information characteristics, the basic attributes (as mentioned above) may be used as indicators for identifying and acknowledging gaps within available information for consideration as it is used. The relevancy aspect of available information must never be compromised because information ought to be sensible to the decision problem in place (Malhotra, 2005).

Based on the prevailing scenario, different techniques are used to support decision-makers (Malhotra, 2005). Different decision-makers normally have dissimilar preferences regarding the nature of the information that they need and their preferred procedure of executing the process, which makes dialogue between individuals significant. As such, decision-makers should share their expectations, objectives, and information gaps, or else the remaining teams may not work efficiently. Thus, they would lose focus or pay attention to things that would be equally detrimental and time-consuming (Malhotra, 2005). Finally, the above unveiled that the decision framework is significant, and it should be used by the decision-makers' preferences. More so, with the scarce nature of time, decision-makers ought to prioritize what they want for effective decision-making based on their previous experience.

2.6 *Challenging Information and Analysis*

Immediately after the analysis is over and information is secured, it is important to critique its quality and relevancy. One of the proposed approaches to doing such is through subjecting it to discussions with other parties involved in decision-making. Through discussions and disagreements, people imagine what compels others to understand the rationale behind contrary opinions (Drucker, 1967). Equally, having conflicting views usually helps in ensuring various opinions are subjected to critique; facts are challenged, and analysis is equally scrutinized before decisions are made. Besides, Shenhar and Levy (2007) reveal that a carefully chosen decision team has a higher probability of ensuring that prevailing assumptions are further challenged, that options are intensively debated, and that biases are revealed while upholding information quality.

Hence, how does decision-making arrive at a consensus on whether a given issue is worthy? Under what circumstance may such an issue become a "decision problem"? Based on Grünig and Kuhn (2009), such appears among the initial steps to be covered to determine whether given decision problem merit to be subjected to a formal process or not. Drucker (1967) asserts that the best approach is by considering how weighty the conflict is between the current and future desired state, and what could be the risk of action versus inaction. If the conflict is established to be substantial and happens to deviate from the desired state, it is wise to take up the process.

Equally, while using the formal process, the risk of "paralysis by analysis" ought not to be forgotten, so the decision problem and options ought to be analyzed many times to the extent that the ultimate decision keeps on to are postponed. Such a path may be challenging, as making decisions requires courage and judgment (Drucker, 1967). In some situations, the decision-maker decides if the maturity point in regard to analysis has been attained. Mostly, this may not be an obvious phase within a fast-changing project environment, but most decisions are arrived at under uncertainty. As such, the decision-maker is mostly guided by their judgment and experience to arrive at the final decision. At times, compromises ought to be made, while the decision-maker should have clear rules on what may be acceptable, so decisions are made based on the available information. After fulfilling such, it is worth considering what should follow.

2.7 Post-decision Activities

Even before imagining what would happen after making a decision, it is prudent to discuss the relevancy of the subject matter. Others assume that what happens after a decision is made is normally out of scope in the decision-making framework. After all, the process ought to be concluded immediately when a decision is made. Nonetheless, for project managers and corporate leaders to achieve desired results and effectiveness, embracing post-decision activities is very significant. Thus, Drucker (1967) reveals: "Unless a decision has 'degenerated into work,' it is not a decision; it is at best a good intention." Further, he asserts that the act of arriving at a decision is usually the most time-consuming part of the process, but equally the most significant. All decisions ought to be translated as close to the working level as possible and more simply (Drucker, 1967; Sutherland, 2004; Tiainen, 2014). Such can only be attained through establishing clear action steps and by answering this question: "What is going to happen now that the decision has been taken?"

Accordingly, Howard (1988) justifies the significance of the same study: "Good decisions can lead to bad outcomes and vice versa." If there is no follow-up, it becomes difficult for one to establish if any good came out of the decision since even the most logical and consistent decision does not guarantee desired outcomes. Additionally, Howard (1988) encourages the need to distinguish between the outcome and the decision for purposes of remembering. Equally, it helps in improving the quality of action when it becomes apparent that the outcome can never be predicted, but through good action, it may be affected. Below are the ways for ensuring proper actions are equally agreed on and taken.

A decision's effectiveness may be evaluated by measuring whether such a decision brought about a difference in terms of money, time, impact, or action. Drucker (1967) equally asserts that an ideal decision ought to comprise of the following components: accountability aspect, deadline outlined, those affected directly clearly named, names of those indirectly affected, what ought to be done after, and those charged with responsibility. This serves as an ideal checklist that ensures all concerned stakeholders are involved in making necessary resolutions. Equally, it considers the elements of formulating decisions, such as framing the rationale on which the decision was made, how the conclusion was made, and by defining the commitment to action. This enables the decision-makers to understand if the decision was good, and it is potential for leading to action. Though the post-decision-making phase is rarely discussed, those who discuss it come to acknowledge its relevance in the decision-making process. Also, researchers emphasize the significance of monitoring and controlling for re-evaluation purposes to ensure that everything is running smoothly. The emphasis is on the decision-maker to have an interest in the outcome of the decision and to have a team to establish how expectations turn out. By following this route, decisions will have a bearing.

2.8 Summary

Through the literature review, several theories in line with the study have been identified that comprise of both decision theories and analytical models. Furthermore, these theories enhance a holistic analysis of various aspects of a given decision problem. With the ever-changing environment and sustainability project environment, there is little time to extensively analyze prevailing situations within organizations before effective decisions are made. Thus, the theories present a structural approach to decision-making mechanisms that may steer project managers and corporate leaders into achieving organizational set objectives through the environmental and sustainability projects they implement.

Thus, the literature review enhanced the researcher to identify the best decision-making approach as an ideal model (Grünig & Kuhn, 2009; Karlsson, 2012; Obi, 2014). The approach proposed in the study covers some of the key aspects that ought to be considered for an effective decision-making process. Nevertheless, other discussed theories equally apply to different situations, as revealed in the literature review. Additionally, features of information and analysis have equally been debated upon, as they play a significant role in the decision-making process. Above everything, information ought to be relevant for the issues at hand that warrant a decision. Information and analysis could be subjected to criticism during team discussions to guarantee quality decisions.

In regard to post-decision activities, the issue of monitoring and tracking appears from time to time in literature. The project management model equally reveals how monitoring and the control of processes are significant for the timely evaluation of results for adjustments' sake (Project Management Institute, 2013). For instance, Drucker (1967) reveals the purpose of action when he states, "Without action, a decision is at its best a good intention." He further emphasized the significance of what takes place once decisions have been made. In this case, it is viewed as one of the most significant aspects of the study, since real change occurs once a decision is made. Thus, it is important to establish if the process creates a worthwhile action, so a thorough follow-up ought to be in place. As such, Drucker (1967); Howard (1988) outlined checkpoints to ensure that this is actualized.

3 Research Methodology

This chapter expounds on the research methodology in the study. The main components include the research design, study population, sampling technique, data collection, and the data analysis plan employed by the researcher.

The research design refers to a plan and structure for a study as conceived by the researcher to obtain the research question. Thus, such a plan refers to the overall scheme that includes an outline of what the investigator intends to do: the hypotheses formulation, the operationalizing study variable, and the final analysis of data (Cooper

& Schindler, 2008). Also, this study adopted a descriptive research design. The approach is based on information that the researcher gathers through questionnaires, interviews, observation, inventories, and rating scales.

3.1 Study Population and Sampling Technique

Saunders, Lewis, and Thornhill (2009) assert that a study population refers to a set of the case from which a study sample is derived. In most cases, such may constitute human beings or not. Equally, a study population refers to a group of people or items that have the same characteristics from which research data are obtained for analysis. Hence, the study population comprised of 5 corporate organizations undertaking a variety of projects. Three senior project managers were chosen using simple random sampling from every organization, which translated into a total of 15 respondents and five directors of the sampled organizations. Hence, the total sample was 20 respondents.

3.2 Data Sources

The study utilized primary data and secondary data. Primary data for the study were collected using questionnaires and interviews with the informants. The organizational directors and senior project managers of sampled firms gave information. Further, secondary data for the study were obtained through an online review of books, journals, articles, magazines, and publications from other reliable sources for research topic information and information of other related studies within different project environments.

3.3 Data Collection

The study used a questionnaire designed on a 5-point scale that ranged from "strongly agree" to "strongly disagree." Thus, both closed-ended and open-ended questionnaires were used. Besides, there was an interview guide that the researcher used to interview the directors of firms that took part in the study. Further, content validity and reliability for the research instrument was ascertained as a set of draft questionnaire to three senior project managers. These managers are involved in day-to-day decision-making for the projects they lead, while four questionnaires were given to other experienced researchers in the field of decision-making. The two groups reviewed the questionnaire content and affirmed that the content items were ideal for gathering relevant data for the study.

3.4 Data Analysis Plan

The study's primary and secondary data were analyzed qualitatively. Data were checked to ensure that all questionnaires had been filled, coded, and edited to ensure that the study had accurate information for analysis. The study questionnaire was comprised of two sections: A and B. Section A sought for demographic data. In contrast, section B sought responses on issues that inform effective economic decision-making in project environments and project life cycle.

4 Findings

The leadership and top management of the five organizations were comprised of directors and below them were senior management teams. All were recruited professionally and competitively. Decision-making in the five organizations was structurally done, and consultations, discussions, and critique formed the process of decision-making.

The study established that the organizations had strategic plans in place that informed the kind of environmental and sustainability projects they undertook. As such, the directors with senior project managers were guided by established strategic plans that informed them of progress, challenges, and areas that needed improvement before decisions were made.

The performance of the organizations was above average in their respective projects, as there were strategies that informed the decision-making process. Consultative meetings took place where directors and senior managers discussed projects about the project environments before making decisions and discussing a way forward for the organizations. The study equally revealed that at one point during the early stages of the organizations, they lacked appropriate decision-making structures that informed their work. As such, the management team was not competitively recruited, which almost made the organizations close.

It was further revealed that decision-making is never an easy task within organizations because of the complex nature of the business environment and the effects of a globalized world. There is a heightened demand for making speedy decisions without having adequate time to discuss, reflect, evaluate, and critique various decision proposals within organizations. As such, the study revealed that directors and senior management teams are faced with a need to make complex decisions within a similar time frame, the need to make decisions amidst uncertainties, and the need for making complex decisions under perceptual decisions in their respective organizations.

Further, the study revealed that organizations are equally compelled to solicit the services of decision-making consultants if the management teams lack sufficient time and expertise to arrive at decisions regarding complex projects. Such consultants then table their recommendations and defend their stance on their recommendations for management teams to consider. The study also established that the project environment significantly influences the approach and nature of decisions from top management in organizations, including the internal and external environments in which projects are undertaken. For effective decision-making, organizations were compelled to invest in information analysis. This included ensuring relevant, sufficient, and up-to-date information to inform the decision-making process. The organizations invested in post-decision activities as follow-up mechanisms. They included monitoring to find if projects are being undertaken according to decisions made and if things go contrary to plan, alternative strategies, and if decisions may be taken immediately.

5 Discussion

Key lessons can be learned from this research. Organizations ought to have a competent team of professionals in the form of directors and managers who oversee the planning, execution, and actualization of planned environmental and sustainability projects. Such teams would give direction in regard to day-to-day management and project implementation. Thus, organizations that have had the wrong people occupying key positions eventually close due to a crisis regarding their decisions. On the other hand, organizations that have qualified and competitively recruited staff have always followed the right procedures of decision-making, so they succeed in implementing their projects in any challenging project environment.

Environmental and sustainability strategic planning within organizations is essential. For instance, having a poor strategy in place affects the functionality of organizations through the projects they implement. Such would lead organizations to incur losses, lose customers, and close, so firms need an up-to-date strategy trade in the right direction. Directors and the senior management team work hand-in-hand while consulting on several aspects before generating decisions that benefit their projects. For example, in such organizations, everything is done professionally, right from planning, execution, and monitoring projects. Systems can determine that departmental performance is in place, as sufficient project data are available for scrutiny, and interdepartmental consultations that enhance effective decision-making are encouraged. Additionally, poor decision-making and planning for finances within organizations affect project implementation negatively. Such would lead to organizations not realizing set project objectives, so a holistic plan would entail aspects of programmatic interventions: resources and finances necessary to enhance an organization attain their goals.

Performance checks are crucial attributes within organizations implementing different projects. Regular monitoring of project progress, reporting on progress, realizing challenges, and finding areas for improvement in subsequent project phases allow organizations to re-examine their strategy. Such is made possible through holding quarterly review meetings of project progress, in which effective decisions are made that may necessitate a change of approach. However, the prevailing approach would be upheld if it works, and approach measures would be crafted to overcome challenges. Teamwork takes priority, as well as consulting from time to time and making effective joint decisions.

Additionally, the study established that decision-making happens to be a critical assignment for executives within organizations. Besides, it can be the riskiest venture because making bad decisions may damage an organization, business entity, or the careers of those within organizations. The study also revealed that bad decisions arise from how organizations make their decisions, as alternatives may not have been established. Also, information may not have been accurate, and costs and benefits could not have been established accurately. Nevertheless, the fault could be not with the decision-making process, but rather in the minds of the decision-makers, as the human brain could sabotage the decisions.

Inadequate or lack of decision-making structures and strategy, especially at the early stages of organizations, could heavily affect the performance and goals realization within organizations. Such challenges may be traced back from the personnel recruitment stage where the non-qualified staff is recruited without adequate experience to run organizational programs. Mostly, young organizations may need the services of decision-making consultants with skills and vast experience to guide them, to make recommendations, and to know what venture to avoid. Thus, organizations would improve and would begin registering enhanced performance in their respective fields.

Moreover, decision-making is such a demanding venture within organizations. The discourse is complex based on the dynamic attributes of the business environment and the teething effects of globalization. There are situations within organizations where decisions have to be made speedily without adequate time for reflection, discussion, and evaluation. Therefore, without adequate skills and expertise, managers can make mistakes with decisions, as accuracy needs to be guaranteed. It was revealed that effective decision-making in organizations emanates from a systemic process, asset within organizations, so firms attain effective decision-making by following a step-by-step process:

- Enterprise Governance: Organizational boards provide the ultimate governance for their respective organizations. They play effective oversight in regard to performance aspects and conformance issues in the organization. Through the formal planning process, firms come to understand their strategic context, brand values, and budgetary checks that inform the decision-making process.
- Context/Mindset: Effective decisions within organizations are made in the context of their strategic direction, ethics, and culture of the individuals who harbor prejudices and attitudes regarding the issue over which they ventilate. It is rare to underestimate people's prejudices over organizational culture, attitudes, and behaviors. Thus, in such circumstances, corporate leaders and project managers ensure that they consider alternatives properly and that the decision-making process is evidence-based.
- Frame the Issue: This is a very significant aspect of decision-making. Issues for deliberation ought to be well framed to balance the broad view altogether

with efficient focus. Appropriate stakeholders should be involved. Moreover, the interests of stakeholders are considered to determine objectives at this level. For profit-oriented enterprises, the key aspect becomes the effect on the shareholder value.
- Assemble Information: This entails providing insightful information that gives a clear picture of the enterprise's prevailing financial and competitive position. Moreover, information is gathered for business proposals, with a value attached to the firm's customers and the effect on the firm's value chain. As such, risks involved need close partnering with the enterprise.
- Select Alternatives: Alternatives should be created based on available evidence and analysis, rather than on individual opinions. Risks should be identified either as deal-breakers or as issues that ought to be managed. Corporate leaders and project managers are better placed to facilitate unbiased and evidence-based decision-making within their respective organizations. As such, they are better placed in providing a consistent qualitative and quantitative analyses of the prevailing situation to make proposals for decisions that could propel their organizations forward.
- Decisions: Those who make decisions ought to have the authority to make decisions. Role clarity and understanding are significant for decisions to be made without delay or being swayed by the interests of other parties.
- Manage Implementation and Impact: To manage implementation through to impact, clear communication of decisions, and anticipated outcomes, as reflected in the organization's performance management metrics, are needed. Thus, the aspect of quantifying or describing likely outcomes enhances implementation to be managed and equally suitable action to be taken promptly. Such ensures that set objectives are realized.
- Feedback: Trial and error could be permitted as tactical experiments falling within tolerable risk parameters, but a repeat of past mistakes ought to be inexcusable. Moreover, the decision and matters under consideration ought to be documented for post audit or learning purposes. As such, outcomes of previous decisions ought to be documented as part of corporate memory for ensuring that lessons are learned in the future.

Therefore, various multinational corporations have managed to transform their project management functions to aid the business by improving decision-making. Such functions include systems, people, processes, and structures that offer timely and accurate management information. Project managers are literate business-wise, as they operate within a culture that appreciates the contributions they make toward evidence-based effective decision-making.

To improve a project of whatever nature is to enhance effective managerial decision-making, as they determine the firm's shared vision. Such a vision ought to be developed and communicated to organizational members by the head to enhance decision-making. Thus, a change agenda ought to consider both effectiveness and efficiency aspects within an organization regarding systems, people, structure, and processes.

5.1 Organizational Implications

Research on the acquired skill and management strategies has revealed that for conducting business projects and project management, these variables, their concepts, and models are necessary and more valuable than technology. The research results emphasize that strategic planning is the most vital aspect of leadership. A top-down and bottom-up approach only leads to temporary solutions to problems, especially with elements of project management, operations management, and process improvement. Furthermore, the results show that leadership styles and tools require these variables, their concepts, and models.

Also, a business' leadership and management need to be trained sufficiently for project management and performance. Within this study, it is revealed that current organizational issues are derived from poor leadership because the bottom-line approach can lead to problems. If project management and operational performance are successfully supervised with the right tools and information, then the performance, profits, and costs of a business will also improve.

This study finds that these variables, their concepts, and models element are overlooked too frequently in favor of financial elements, which only affect the short term. To have a good long-term strategy for leadership, it is important to emphasize many different aspects of a business. Operations, project management, financials, performance, strategy, and human resources are all equally important in any organization. This perspective will bring about success in the short term and the long term.

5.2 Managerial and Team Implications

Primarily, the results examine the variables, concepts, and models to fill a gap in knowledge within all other research. The performance and effectiveness of an organization can be affected by these variables, concepts, and models, which make their relationships vital.

This study also features an outline for projects and performances, as knowing the relationships of the variables, concepts, and models yield better management. As a result, leaders can generate more inclusive mentoring or managerial constructs for teams and businesses to find shortcomings and why they occur. Performance gaps will become more detectable for teams, as well, so that teams can use the tools to benefit a project or departmental goals.

The final implication of this study is that it reveals the benefits of more inclusive training programs. With this study, businesses can be guided to improve the team and organizational performance and effectiveness, especially for project teams, project leadership, and organizational leadership. Such training can show how to measure a team, project, or business' performance up against standard and industry-accepted models and concepts. Also, this study contains useful information for leaders to manage teams and projects by educating teams and leaders on how their relationship impacts the business. Businesses will improve their performance and effectiveness.

5.3 Implications and Applications to Fields of Project Management and Engineering Management

Environmental and sustainability decision-making is increasingly the basis for value creation and competitive advantage within contemporary organizations. Moreover, organizations with enhanced decision-making structures ultimately realize that superior business performance if global markets enhance firms to access equal resources. Additionally, competition leads to various enterprising processes to meet on world-class standards. As such, the quality of decision-making acts as a key differentiator that links in a firm's value chain.

Through effective decision-making, various organizations are better placed to formalize their environmental and sustainability strategic planning processes and their governance structures. The planning process is ideal in reports generation, which eventually determines the kind of decisions undertaken by organizations. Effective decision-making equally helps to formulate policies, implementing strategic plans that are vital for strategic management, vision, missions, and objectives realization. Thus, plans are implemented and actualized through programs that transform livelihoods or beneficiaries. Effective decision-making is critical to approach challenges in a professional manner. For instance, issues related to the competition are tackled professionally while following laid strategies. Effective decision-making equally helps firms to learn from their previous mistakes to arrive at decisions that steer organizations forward. In turn, previous decisions become reference points for future decisions.

Though environmental and sustainability projects require these variables, their concepts, and models, the role of the engineers and technical professions is just as important. Engineers used to be required to use technology and mathematical tools for problem-solving, but they now must use these tools for problem-solving for economically viable solutions. Thus, these variables, their concepts, and models are important players in engineering decisions, as economically sound products will lead to good profits and performances. Engineers should also be aware of business management and maturity models so that their knowledge can benefit their investors.

Management and engineering are interrelated, as management and engineering concepts are both scientific. For instance, engineering relies on the cause and effect relationship, which is considered scientific, so engineering has relied on management to improve projects. Though research does address that the models can identify project elements, this is explained through a business perspective. This study aims to take an engineering perspective, as well as pure engineering filed techniques, such as budgeting, equipment, and purchasing material. Engineers and project managers can find many decision-making methods to address engineering problems, as well as screening projects for viability.

This research relies on scholarly information on these variables, their concepts, and models to analyze them and to find their influence on project management and operational performance. Thus, this study will find the best practices for these variables, their concepts, and models to provide a future reference for the IE/EM research field. It also features useful information on project management and operational performance. Though these variables, concepts, and models can create a different environment in the IE/EM profession, the structure of a scope can make IE/EM players generate the required scopes of interest at any level, so the development of these variables is applied with the given scopes.

Furthermore, this research can contribute to stakeholders (system engineers, project managers, etc.) for applying maturity to project management. In business, each field requires sufficient project management, as new products and services must be relevant. Also, stakeholders can maximize on the roles of system engineering and project management for greater success in business projects. Effective decision-making theories discussed in the study may guide corporate leaders and managers in their decision-making structures within their respective organizations. Applying the theories in unique situations that warrant decision-making would make this happen. Thus, the relationships between these variables, their concepts, and models are of great importance.

Though much literature addresses these variables, their concepts, and models, this research covers new ground by studying how they affect project management and operational performance development. Systems thinking logic, along with new product development objectives, can lead products to market profitably. With this study, researchers can see ways for a small company to create new products, even without many established processes; this is only the second product from the company under consideration in this study.

6 Conclusions

Ultimately, the study concludes that the environmental and sustainability decision-making process entails making choices from a wide pool of alternatives. Decisions are equally commitments to taking action. Each decision is tricky, so this entails committing available resources to a certain and uncertain future. Experienced project managers and corporate executives can detect roadblocks that could hinder effective decision-making, which will develop strategies to overcome them. Thus, effective decision-making calls for a precise strategy that produces desired results.

Organizations have both strategic and non-strategic sub-categories of environmental and sustainability decisions. Strategic decisions determine an organization's ultimate direction. On the other hand, non-strategic decisions refer to the organizational day-by-day minor operation-based decisions. As such, effective decision-making necessitates precise strategies to produce desired outcomes, such as brainstorming, bargaining, trial and error, and nominal grouping.

Moreover, other hidden traps that may be avoided in the organizational decision-making process include being over-confident, framing, status-quo trap, anchor trap, and sunk-cost trap. Hence, decision-making is never an easy venture, more so with managers running contemporary organizations. With intense organizational changes regarding project environments, the art of making decisions is equally accelerating. So, for managers to cope with such change, decision-makers are faced with challenging situations as they venture into making decisions. Such challenges may include situations where complex streams of decisions are to be made and equally arriving at such decisions in the wake of uncertainties.

6.1 Recommendations for Future Research

The study took a general perspective of effective decision-making by corporate leaders and project managers. Thus, more research is needed in the decision-making field, specifically on the influence of culture on effective decision-making, as it impacts project management and implementation in organizations. Such would help in a critical understanding of the decision-making concept within an organizational framework.

Further, more research should be conducted on the effect of the internal project environment on decision-making for project managers to accept the best environment. Additionally, considering the limited nature of respondents that took part in the study and the approach used, it is prudent that future research is done with an increased number of respondents and firms to gain ground for generalizing results. In terms of approach, the researcher recommends that both qualitative and quantitative approaches be used during the proposed study for efficiency and future results.

Future research could also assess these factors within other industries and managerial settings to find their strengths, weaknesses, and what impacts them. Research can also find how these variables, concepts, models, and their relationships are perceived in organizational, strategic, or cultural viewpoints. Thus, research can find how culture, strategy, human resources, and operations affect these variables, concepts, and models.

6.2 Limitations

Since this study is designated to research, the study design faced the following challenges. The study was limited to five firms and interacted with the executives and senior management teams. Thus, there was a limited sample size, which could lead to bias and validity within the findings and conclusions. Larger sample size would have been preferable, as this smaller size limited the participation of other firms that were willing to take part in the research. Due to the limited nature of time, only a few participants took part and were interviewed. This limited the information base, as the lower cadre managers within organizations bear critical information regarding the decision-making process within organizations. Moreover, having a limited number of sample sizes prove difficult to generalize the study outcome over the population size.

Also, the fact that the research mainly centered on project managers and executives could be regarded as a limitation, as no consideration was given to projects and project management. This study only assessed the key factors and their relationship from a project environment, so the conclusions and analysis only apply to project environments. The findings may not apply to other areas, either (i.e., supply chain management, operations management, or strategic management). Thus, this study cannot assert that the findings can be deployed in other industries or managerial settings.

Ultimately, it is worth noting that there were those limitations that were beyond control, especially in the administration of interviews. As a response, the researcher had mechanisms that helped in attaining desired responses and in the interview administration.

6.3 *Final Thoughts*

The study concludes that decision-making is a very critical process for project managers and corporate leaders, as it lays the basis for environmental and sustainability strategic planning within organizations. For a sufficient decision-making process, organizations should take the best practice approach as a guideline to their decision-making process. This includes timely consultations, availing relevant information on time, and critiquing available alternatives based on facts. Organizational strategic management practices' success entirely depends on decision-making processes. For instance, project planning, implementing, oversight strategy, and performance strategy rely on the nature of decision-making within organizations.

References

Ahern, T., Leavy, B., & Byrne, P. J. (2014). Complex project management as complex problem solving: A distributed knowledge management perspective. *International Journal of Project Management, 32*(8), 1371–1381.

Albert, C. (2006). Decision making. *Harvard Business Review, January*.

Al-Kadeem, R., Backar, S., Eldardiry, M., & Haddad, H. (2017). Review of using system dynamics in designing work systems of project organizations: Product development process case study. *International Journal of System Dynamics Applications (IJSDA), 6*(2), 52–70. https://doi.org/10.4018/IJSDA.2017040103.

Andersen, E. S. (2014). Value creation using the mission breakdown structure. *International Journal of Project Management, 32*, 885–892.

Arumugam, V. A. (2016). The influence of challenging goals and structured method on six sigma project performance: A mediated moderation analysis. *European Journal of Operational Research, 254*(1), 202–213.

Badi, S. M., & Pryke, S. (2016). Assessing the impact of risk allocation on sustainable energy innovation (SEI): The case of private finance initiative (PFI) school projects. *International Journal of Managing Projects in Business, 9*(2), 259–281.

Besner, C., & Hobbs B. (2012). The paradox of risk management: A project management practice perspective. *International Journal of Managing Projects in Business, 5*(2), 230–247.

Brown, S. L., & Eisenhardt, K. M. (1995). Product development: Past research, present findings, and future directions. *Academy of Management Review, 20*(2), 343–378.

Buchanan, L., & O Connell, A. (2006). A brief history of decision-making. *Harvard Business Review, 84*(1), 32.

Burnes, B. (2014). Kurt Lewin and the planned approach to change: A re-appraisal. *Journal of Management Studies, 41*(6), 977–1002.

Campbell, A., Whitehead, J., & Finkelstein, S. (2009). Why good leaders make bad decisions. *Harvard Business Review, 87*(2), 60–66.

Cooper, J., & Schindler, M. (2008). *Perfect sample size in research*. New Jersey.

Cova, B., & Salle, R. (2005). Six key points to merge project marketing into project management. *International Journal of Project Management, 23*(5), 354–359.

David, M. E., David, F. R., & David, F. R. (2017). The quantitative strategic planning matrix: A new marketing tool. *Journal of Strategic Marketing, 25*(4), 342–352.

Detert, J. R. (2000). A framework for linking culture and improvement initiatives in organizations. *Academy of Management Review, 25*(4), 850–863.

Drucker, P. (1967). *The effective executive. 1*, Collins.

Easton, G. S., & Rosenzweig, E. D. (2012). The role of experience in six sigma project success: An empirical analysis of improvement projects. *Journal of Operations Management, 30*(7), 481–493.

Eskerod, P., & Blichfeldt, B. S. (2005). Managing team entrees and withdrawals during the project life cycle. *International Journal of Project Management, 23*(7), 495–503.

Galli, B. (2018a). Application of system engineering to project management: How to view their relationship. *International Journal of System Dynamics Applications, 7*(4), 76–97.

Galli, B. (2018b). Can project management help improve lean six sigma? *IEEE Engineering Management Review, 46*(2), 55–64.

Galli, B. (2018c). Risks related to lean six sigma deployment and sustainment risks: How project management can help. *International Journal of Service Science, Management, Engineering, & Technology, 9*(3), 82–105.

Galli, B., & Hernandez-Lopez, P. (2018). Risks management in agile new product development project environments: A review of literature. *International Journal of Risk & Contingency Management (IJRCM), 7*(4), 37–67.

Galli, B., & Kaviani, M. A. (2018). The impacts of risk on deploying and sustaining lean Six sigma initiatives. *International Journal of Risk & Contingency Management, 7*(1), 46–70.

Galli, B., Kaviani, M. A., Bottani, E., & Murino, T. (2017). An investigation of shared leadership & key performance indicators in six sigma projects. *International Journal of Strategic Decision Sciences (IJSDS), 8*(4), 1–45.

Gänswein, W. (2011). *Effectiveness of information use for strategic decision making* (*entrepreneurship*). Gabler Verlag.

Goodland, R. (2005). The concept of environmental sustainability. *Annual Review of Ecology and Systematics, 26*(2), 1–24.

Gimenez-Espin, J. A.-J.-C. (2013). Organizational culture for total quality management. *Total Quality Management & Business Excellence, 24*(5–6), 678–692.

Grünig, R., & Kühn, R. (2009). *Successful decision-making: A systematic approach to complex problems* (2nd ed.). Springer.

Hamel, G. (2006). The why what, and how of management innovation. *Harvard Business Review, 84*(2), 72.
Hammond, J. S., Keeney, R. L., & Raiffa, H. (2006). The hidden traps in decision-making. *Harvard Business Review, 84*(1), 118.
Hartono, B., Wijaya, D. F. N., & Arini, M. H. (2014). An empirically verified project risk maturity model: Evidence from Indonesian construction industry. *International Journal of Managing Projects in Business, 7*(2), 263–284.
Hoon Kwak, Y., & Dixon, C. K. (2008). Risk management framework for pharmaceutical research and development projects. *International Journal of Managing Projects in Business, 1*(4), 552–565.
Howard, R. (1988). Decision analysis: Practice and promise. *Management Science, 34*(6), 679–695.
Karlsson, J. (2012). *Decision-making in a multinational manufacturing organization* (Master's thesis). Helsinki: Aalto University School of Business.
Labedz, C. S., & Gray, J. R. (2013). Accounting for lean implementation in government enterprise: Intended and unintended consequences. *International Journal of System Dynamics Applications (IJSDA), 2*(1), 14–36.
Lamming, R., & Hampson, J. (2000). The environment as a supply chain management issue. *British Journal of Management, 7*(S1), S45–S62.
Lee, J., Lapira, E., Bagheri, B., & Kao, H. (2013). Recent advances and trends in predictive manufacturing systems in a big data environment. *Journal of Cleaner Production, 3*(10), 45–55.
Loyd, N. (2016). Implementation of a plan-do-check-act pedagogy in industrial engineering education. *International Journal of Engineering Education, 32*(3), 1260–1267.
Malhotra, N. (2005). *Marketing research. An applied approach-european* (2nd ed.). Financial Times Management.
Mansfield, E. (1999). *Managerial economics* (4th ed.). 500 Fifth Avenue, New York, NY, 10110, U.S.A.: W.W. Norton & Company Inc.
Marcelino-Sádaba, S., Pérez-Ezcurdia, A., Lazcano, A. M. E., & Villanueva, P. (2014). Project risk management methodology for small firms. *International Journal of Project Management, 32*(2), 327–340.
Medina, R., & Medina, A. (2015). The competence loop: Competence management in knowledge-intensive, project-intensive organizations. *International Journal of Managing Projects in Business, 8*(2), 279–299.
Milner, C. D. (2016). Modeling continuous improvement evolution in the service sector: A comparative case study. *International Journal of Quality and Service Sciences, 8*(3), 438–460.
Nagel, R. (2015). Operational optimization: A lean, six sigma approach to sustainability. *Proceedings of the Water Environment Federation, 3*(4), 1–12.
Obi, J. N., & Agwu, M. E. (2017). Effective decision-making and organizational goal achievement in a depressed economy. *International Journal of Research and Development Studies, 8*(1).
Obi, J. N. (2014). Decision-making strategy. In C. P. Maduabum (Ed.), *Contemporary issues on management in organizations of readings* (p. 63). Ibadan: Spectrum Books Limited.
Papke-Shields, K. E., & Boyer-Wright, K. M. (2017). Strategic planning characteristics applied to project management. *International Journal of Project Management, 35*(2), 169–179.
Parast, M. M. (2011). The effect of six sigma projects on innovation and firm performance. *International Journal of Project Management, 29*(1), 45–55.
Parker, D. W., Parsons, N., & Isharyanto, F. (2015). The inclusion of strategic management theories to project management. *International Journal of Managing Projects in Business, 8*(3), 552–573.
Project Management Institute. (2013). A guide to the project management body of knowledge.
Rogers, P., & Blenko, M. (2006, January). Who has the D? How clear decision roles enhance organizational performance. *Harvard Business Review*.
Saaty, T. L. (2008). Decision making with the analytic hierarchy process. *International Journal of Services Sciences, 1*(1), 83–98.
Saunders, M., Lewis, P., & Thornhill, A. (2009). *Research methods for business students*. Pearson Education.

Schwedes, O., Riedel, V., & Dziekan, K. (2017). Project planning vs. strategic planning: Promoting a different perspective for sustainable transport policy in European R&D projects. *Case Studies on Transport Policy, 5*(1), 31–37.
Sharon, A., Weck, O. L., & Dori, D. (2013). Improving project-product lifecycle management with model-based design structure matrix: A joint project management and systems engineering approach. *Systems Engineering, 16*(4), 413–426.
Shenhar, A. J., & Levy, O. (2007). Mapping the dimensions of project success. *Project Management Journal, 28,* 5–13.
Steinburner, T. (1974). Decision-making strategies. *School of Management Studies.*
Sutherland, S. (2004). Creating a culture of data use for continuous improvement: A case study of an Edison project school. *American Journal of Evaluation, 25*(3), 277–293.
Svejvig, P., & Andersen, P. (2015). Rethinking project management: A structured literature review with a critical look at the brave new world. *International Journal of Project Management, 33,* 278–290.
Tiainen, A. (2014). *Decision-making in project management.*
Todorović, M. L., Petrović, D. Č., Mihić, M. M., Obradović, V. L., & Bushuyev, S. D. (2015). Project success analysis framework: A knowledge-based approach in project management. *International Journal of Project Management, 33*(4), 772–783.
Ulbrich, F. (2010). Adopting shared services in a public-sector organization. *Transforming Government: People, Process, and Policy, 4*(3), 249–265.
Usman Tariq, M. (2013). A six sigma based risk management framework for handling undesired effects associated with delays in project completion. *International Journal of Lean Six Sigma, 4*(3), 265–279.
Von Thiele Schwarz, U. N.-H. (2017). Using kaizen to improve employee well-being: Results from two organizational intervention studies. *Human Relations, 70*(8), 966–993.
Wideman, R. M. (1990). Managing the project environment. *Dimensions of project management* (pp. 51–69). Berlin, Heidelberg: Springer.
Winter, M., Andersen, E. S., Elvin, R., & Levene, R. (2006). Focusing on business projects as an area for future research: An exploratory discussion of four different perspectives. *International Journal of Project Management, 24,* 699–709.
Xiong, W., Zhao, X., Yuan, J.-F., & Luo, S. (2017). Ex post risk management in public-private partnership infrastructure projects. *Project Management Journal, 48*(3), 76–89.
Xue, R., Baron, C., & Esteban, P. (2016). Improving cooperation between systems engineers and project managers in engineering projects-towards the alignment of systems engineering and project management standards and guides. *Proceedings of Joint Conference on Mechanical, Design Engineering & Advanced Manufacturing, 24*(2), 23–40.
Xue, R., Baron, C., & Esteban, P. (2017). Optimizing product development in industry by the alignment of the ISO/IEC 15288 systems engineering standard and the PMBoK guide. *International Journal of Product Development, 22*(1), 65–80.
Yun, S., Choi, J., Oliveira, D. P., Mulva, S. P., & Kang, Y. (2016). Measuring project management inputs throughout capital project delivery. *International Journal of Project Management, 34*(7), 1167–1182.
Zhang, X., Bao, H., Wang, H., & Skitmore, M. (2016). A model for determining the optimal project life span and concession period of BOT projects. *International Journal of Project Management, 34*(3), 523–532.
Zwikael, O., & Smyrk, J. (2012). A general framework for gauging the performance of initiatives to enhance organizational value. *British Journal of Management, 23,* S6–S22.

Development of Policies and Practices of Social Responsibility in Portuguese Companies: Implications of the SA8000 Standard

Ana Costa, João Leite Ribeiro, and Delfina Gomes

Abstract This study aims to explore the implications of the SA8000 standard in the implementation of Corporate Social Responsibility (CSR) policies and practices in Portuguese companies. Although this is not a recent issue, it is a subject that has motivated a growing interest on the part of the business and academic community. CSR should be seen as a new business management model, which includes universal human values as well as ethical decisions that ensure the satisfaction of the interests and needs of all stakeholders and the community in general. Regarding the empirical aspect of this research, a qualitative approach was followed through the application of a semi-structured interview to certified and not certified companies by SA8000 standard, in order to understand possible disparities with regard to the implementation of CSR policies and practices. Although the certification is the guarantee of the commitment of the companies with the CSR, the truth is that this is seen by the companies as a form of differentiation in the market. In general terms, it can be concluded that the policies and practices of CSR do not differ in relation to certification, which means that all companies have the possibility of making a difference by being socially responsible.

Keywords Portuguese companies · Policies and practices of Corporate Social Responsibility · Corporate Social Responsibility · SA8000

1 Introduction

> We are not only responsible for what we do, but also for what we do not do!
>
> Jean-Baptiste Molière (quote in Gardner, 2006, p. 235)

Over time, management has evolved greatly, not only socially but also culturally, economically, environmentally and technologically, where CSR is also an integral part of management that must be balanced, accountable and sustainable. Although

A. Costa · J. L. Ribeiro (✉) · D. Gomes
School of Economics and Management, University of Minho, Braga, Portugal
e-mail: joser@eeg.uminho.pt

C. Machado and J. P. Davim (eds.), *Circular Economy and Engineering*, Management and Industrial Engineering,
https://doi.org/10.1007/978-3-030-43044-3_3

CSR is not a recent issue, it is a subject that has generated a great deal of interest not only in business but also in academia, and there is still no definition of universal acceptance. For several years, CSR was mixed up with philanthropy, which had nothing to do with the central object of the business. In fact, these actions may be included in the company's CSR, but this is not what makes it socially responsible.

CSR should also not be seen as an innovative marketing mechanism, but as a new management behavior model, which includes universal human values and ethical decisions, which ensure the interests of all stakeholders and the community at large. The constant involvement of stakeholders in the implementation of CSR practices allows companies to reflect when they avoid following the standard of socially acceptable CSR (Stigzelius & Mark-Herbert, 2009).

Stakeholders are emerging as one of the main reasons for the increased importance given to CSR by their pressure on organizations. By way of example, it can be highlighted investors' preference for socially responsible investments, the growth of responsible consumption, the greater concern on the part of international institutions, private or public and the interest of future employees in being part of companies with CSR practices (Comissão Europeia, 2002).

Although, in recent decades, the growth of the world economy has generated some lack of concern for the environment, nowadays, the panorama has changed, with a growing concern with sustainable development. Thus, many companies have sought to manage processes in a way that combines economic growth and increased competitiveness with preserving the environment and promoting ethical and socially responsible behavior, meeting the needs and interests of society. Therefore, companies wishing to adopt a socially responsible attitude should regard their mission as a tool for development and wealth generation, not solely as a source of profit (Duarte, Mouro, & Neves, 2010; Jorge & Silva, 2011; Azevedo, Ende, & Wittmann, 2016).

Although it is a long-term investment, there are organizations that value CSR, although they are not certified, which is interesting for this research, because it allows the understanding of the practices that are carried out by certified and uncertified companies. But, it is essential to take into account the Portuguese context, which has been trying to solve a socioeconomic crisis, which has resulted in an increase in inequalities and social concerns. The research underlying this study aims to understand the implications of the SA8000 certification standard in the implementation of socially responsible policies and practices in Portuguese companies. The underlying research questions are the following:

- Does SA8000 certification influence corporate-led CSR practices?
- Is there full involvement of all stakeholders in the certification process?
- Do the benefits of CSR practices outweigh the investment?

This chapter is structured as follows. The next section presents a review of the literature beginning with the concept of CSR and proceeding with the dimensions and practices of CSR and concluding with the SA8000 Standard. The context and research method of the study are presented in the subsequent section. Next, the findings of the study are described and analyzed focusing on companies contextualization, implementation of CSR policies and practices and the SA8000 certification process. The final section offers the final remarks.

2 Literature Review

2.1 Concept of Corporate Social Responsibility

Over time, the concept of CSR has been the target of many perspectives that make it a very complex concept. From the perspective of the Commission of the European Communities (CEC) (2001), CSR is a voluntary way for companies to contribute to a fairer society and a more caring environment. This voluntary behavior implies a coherent interconnection between socio-environmental concerns and the business strategy defined by the company. For other authors, CSR is seen as an effort to improve community well-being, which presupposes discretionary business practices and corporate resource contribution (Kotler & Lee, 2005; Costa, 2005).

CSR is a hot topic today, as it is a new management model, not a fad, which may require restructuring and changing mindsets. In addition, it should not be viewed as a business, but as a consistent practice of commitment to the community, that does not merely hide corporate misconduct (Rodrigues, Seabra, & Ramalho, 2009).

This concept has become more visible due to factors such as the opening of trade borders, the global market and the growing technological development. However, it is still often confused with the concept of philanthropy. The difference between these concepts is related to the periodicity of the actions, since in philanthropy, they are casual, while in the case of CSR, we speak of continuous actions, framed in the company's strategy, which converge to value creation and social development (Correia, 2013). It can then be seen that companies are viewed as a social entity that interacts with all socioeconomic agents and have a set of rights and duties that go beyond the legal obligations in force in the legal framework.

More recently, the Ethos Institute (2017), as the Organization for Economic Co-operation and Development (OECD) (2004) had already done, defined CSR as a form of management based on the company's ethical and transparent relationship with the public with who relates. It also involves setting business goals, along with the sustainable development of society, protecting environmental and cultural resources for future generations, respecting and encouraging the reduction of social inequalities. Carroll (1979, 2016) argues that CSR understands the economic, legal, ethical and philanthropic expectations that a society has about organizations at any given time. To explain this concept, Carroll (1979, 2016) defined four categories, framed in Fig. 1.

Economic responsibility is the foundation of the pyramid because it is the basis for the operation of any business. The company must be profitable, sell its products lucratively and maximize its competitive position. In the absence of economic responsibilities, all other responsibilities are affected.

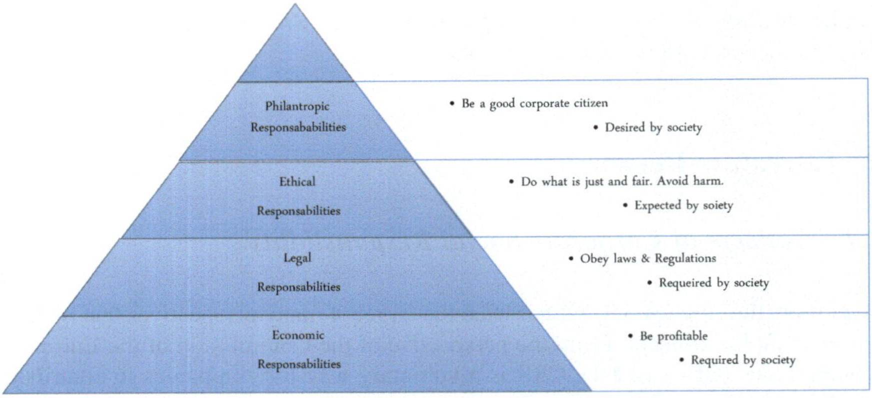

Fig. 1 Pyramid of corporate social responsibility. *Source* Adapted from Carroll (2016, p. 5)

Like economic responsibilities, legal responsibilities are also imposed on companies, that is, organizations must combine their economic mission with the requirements defined by the legal system of society. The company is expected to be governed by laws and regulations where the final product is the result of compliance with the current safety and environmental standards. The company is also expected to conduct its business with ethical behavior (ethical responsibility) based on respect for human rights and in accordance with the principles of fairness, justice and impartiality (Carroll, 1979; Leandro & Rebelo, 2011; Carroll, 2016).

At the top of the pyramid, Carroll (1979, 2016) places philanthropic responsibilities as desirable in an organization. This last level portrays the voluntary actions carried out by the company, which are not covered by the legal and ethical obligation to which the organization is subject. These activities emerge as a way for the organization to contribute socially. Examples include corporate donations to community projects, the implementation of benefits not only for employees, but also for their families, among others where direct returns to the organization are not expected (Carroll, 1979, 2016).

In another study, Schwartz and Carroll (2003) found that this model could result in a different interpretation than expected. The fact that the pyramid is vertical may lead us to believe that there is no interconnection between the different levels, which does not correspond to reality. Thus, they propose a restructuring of the model, having a circular shape (see Fig. 2), with only three categories, economic, legal and ethical, where all relate, without any supremacy of any.

There are companies that disclose actions that fall under their legal responsibility as CSR actions, not distinguishing these two concepts, with the purpose of obtaining image gains. Therefore, since CSR is based on ethical standards, a company that is governed by instrumental standards cannot be considered as socially responsible (Wood, 1991; Mendonça & Gonçalves, 2004; Rego, Cunha, Costa, Gonçalves, & Cabral-Cardoso, 2007). Responsible organizations are based on Elkington's (2004)

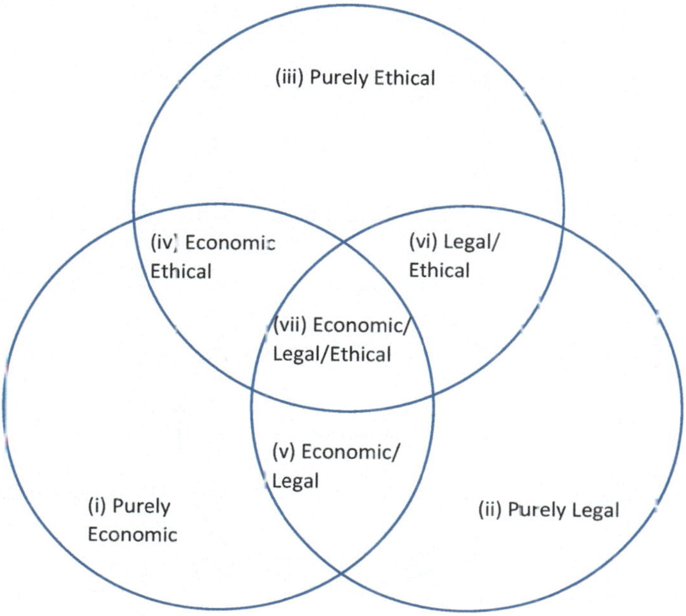

Fig. 2 Three-domain model of corporate social responsibility. *Source* Schwartz and Carroll (2003, p. 504)

"Triple Bottom Line" management model, also known as 3Ps (people, planet and profit).

According to Fig. 3, organizations must be socially and environmentally responsible without losing their economic sustainability. The organization not only aims to meet its own needs, but also seeks to contribute to the well-being of the present and future generations. Sustainability has increasingly been seen as an imperative across any organization (Elkington, 2004).

In Perrini's (2006) perspective, economic, social and environmental issues should be considered in business strategies and company actions, which is a major challenge for business management. Based on this idea, it can be stated that CSR will only be confirmed when it is incorporated into the practices of individuals who work in and with the company that assumes this attitude (Leandro & Rebelo, 2011).

CSR presupposes a continuous, progressive and voluntary process that combines, on the one hand, the organization's ability to collaborate and relate to the community on social and environmental issues, taking into account their values and attitudes to achieve their goals. On the other hand, it involves not only regulatory processes (norms, stakeholder management and business ethics, social marketing and social and ecological labels), but also processes of compliance with environmental and legal obligations (Mascarenhas & Costa, 2011). However, CSR practices are still closely associated with social marketing, due to the media space devoted to them, which has

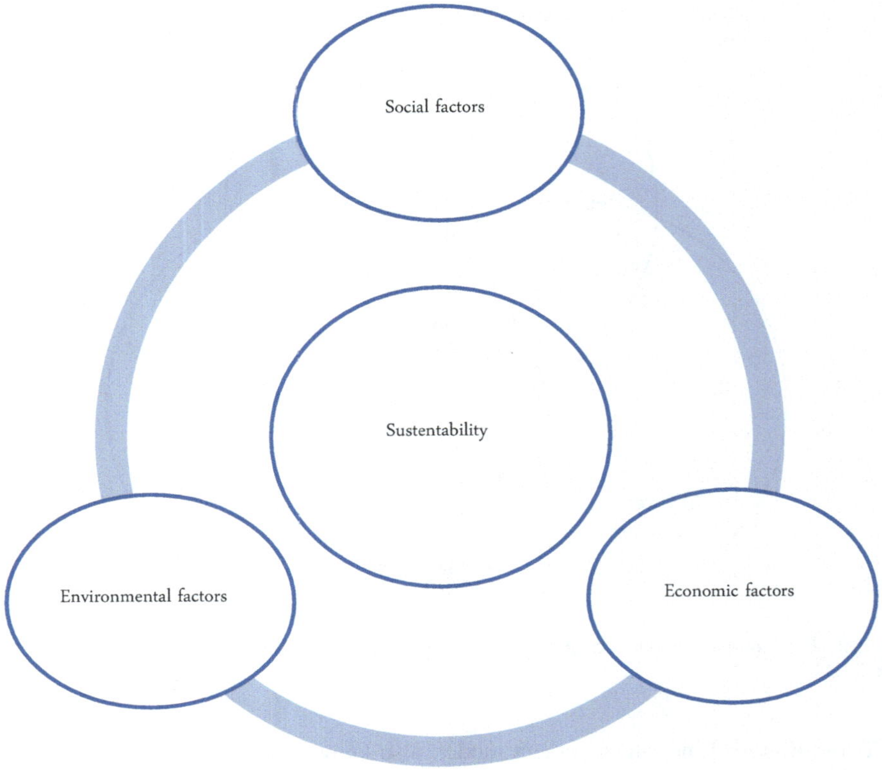

Fig. 3 Model "Triple Bottom Line". *Source* adapted from Elkington (2004)

led to the criticism that socially responsible companies do not act beyond profit, but that is the profit they seek (Leandro & Rebelo, 2011).

When talking about CSR, it is essential to talk about communication. In today's world, it is necessary to take into account the power that is given to each citizen to communicate, and elements such as workers, consumers, competition, social communication and social networks should not be disregarded. Nowadays, movements for or against a particular organization can easily be created from social networks, which can quickly reach large numbers of people (Leal, Caetano, Brandão, Duarte, & Gouveia, 2011).

In summary, it is essential to understand the differences between the concepts of CSR and philanthropy, which relate to the involvement of companies in the implementation of policies and programs that seek to change management practices, not necessarily required by law (Blowfied & Murray, 2008). CSR is a concept that has confirmed a growing importance not only in business but also in academia and should be aligned with sustainable development strategies.

2.2 Dimensions of CSR

Implementing CSR policies produces medium- and long-term results for organizations. This is explained by the fact that if the company is sensitive to the needs of society, it can trigger a better way of guiding business. It can be said, then, that a greater investment in social policies can lead to more profit by the company, and in this way, the company can contribute more effectively to the welfare of society (CEC, 2001).

A socially responsible business needs to adopt behaviors that are not exclusively about complying with legal obligations. The company needs to invest more in human capital, the environment and relationships with other stakeholders and local communities (CEC, 2001; Balonas, 2014). In this way, it is possible to distinguish two dimensions into which CSR practices are divided: the internal dimension and the external dimension. The internal dimension of CSR focuses mainly on four areas: human resource management, occupational health and safety, adaptation to change and environmental and natural resource management (CEC, 2001).

According to the Commission of the European Communities (2001), one of the biggest challenges organizations face is attracting and retaining qualified employees. Therefore, CSR practices should be directed to employees, in order to meet their needs and, consequently, retain the best employees and increase productivity (Balonas, 2014). Thus, the organization needs to strive for the personal and professional development of its employees, the improvement of working conditions and interpersonal relationships.

Therefore, it is essential that companies seek to adopt measures that encourage employee participation in company decisions, continuous learning through training, reconciling work, personal and family life, equal pay and career progression. It is also crucial to highlight measures such as pay equity and career advancement prospects for women and non-discriminatory recruitment practices, which in the latter case will be a way to reduce unemployment rates by encouraging the hiring of older people, older women, women and ethnic minorities (CEC, 2001). Some authors also consider that the implementation of participatory management increases employee engagement, employee identification with company objectives and promotes professional and personal development (Lourenço & Schröder, 2003; Costa, 2005).

With regard to occupational health and safety, the pattern of socially responsible behavior is clearly reflected in the legislative measures. However, with the outsourcing of work currently observed, it is more difficult for companies to be able to control health and safety conditions because they are dependent on their contractors. At this level, certification and product labeling programs as well as management and subcontracting system certification programs have gained importance. With regard to the management of environmental impact and natural resources, the main objective is to reduce the environmental impact caused by industrial processes and distribution of goods and services (CEC, 2001; Azevedo, Ende, & Wittmann, 2016).

Like the internal dimension, the external dimension is also made up of several groups: local communities, business partners, suppliers and consumers, human rights associations and associations with global environmental concerns (CEC, 2001). In this way, the external dimension represents the CSR that crosses the barriers of the company, and involves the local community. The company's interaction with the local community is seen as a crucial agent for the growth of society. This is because companies offer jobs, compensation, benefits and pay taxes, but they are also dependent on the stability and prosperity of their environment. It can be said that greater involvement of organizations with the local community results in greater productivity and competitiveness on the part of the organization (CEC, 2001; Costa, 2005). Therefore, the company can demonstrate its participation in the community through actions such as training, support for local community causes, recruitment of socially excluded people and the design of structures that aim to improve the living conditions of workers' families.

According to the Commission of the European Communities (2001), the relationship of companies with their trading partners and suppliers, while being dynamic, can provide an advantage for companies because it can reduce the complexity and costs of operations. For consumers, the organization must meet their needs in an ethical, efficient and environmentally friendly manner. However, it should be noted that this aspect is somewhat complex as it covers civil, political, economic, social and cultural rights. Community pressure on companies forces them to adopt codes of conduct involving factors such as working conditions, human rights and environmental aspects.

Finally, the Commission of the European Communities (2001) further states that companies must pay attention to global environmental concerns, in particular the impact that the production chain can have on the environment. This is because a large part of environmental problems are related to the activity and exploitation of resources by companies. Therefore, it is important that companies take a socially responsible attitude, aiming to reduce waste production, pollutant emissions and energy consumption.

CSR practices should address both the internal and external dimensions, as they not only benefit all stakeholders, but also enhance value creation and contribute to sustainable development.

2.3 CSR Practices: From Need to Implementation

Increasingly, companies have a clearer notion of their role in sustainable development by developing CSR actions with the aim of contributing to the creation of a responsibility-based culture, which results in sustainable economic, social and environmental development.

Thus, it is essential that companies begin by preparing a mission statement, code of conduct or statement of principles, where they highlight their objectives, core values and responsibilities to stakeholders. Subsequently, they should align these values with their current strategy and decisions. In practical terms, this means including a socially or environmentally responsible dimension in the company's business plans and budgets and evaluating the results by setting up socially oriented advisory committees that conduct social and/or environmental audits or implement training programs (CEC, 2001).

In addition, it is essential that the company develops a social report as a way to measure the environmental and social impact, as well as the quality of its relationship with its stakeholders (Balonas, 2014). This document allows quantifying and qualifying certain indicators that, consequently, will support the company's analysis and decision-making, always based on the results obtained. This document also allows planning, coordination and rationalization of available resources (Rego et al., 2007).

The Commission of the European Communities also points to the need for CSR reporting and auditing as a mechanism for recognizing the company's environmental and social performance. As a way to increase the quality of verification, this entity proposes the involvement of unions and NGOs (CEC, 2001).

Stakeholders have also shown a growing curiosity about accessing company information, enabling them to gain an overall view of the company's behavior, not just its financial performance, as depicted in the financial statements (Rego et al., 2007). It is noteworthy that in SMEs, these reports are more informal and voluntary, while in larger companies, this document is more rigorous and is usually prepared according to the guidelines of the Global Reporting Initiative (GRI) guide (European Commission, 2011; Rego et al., 2007). These guidelines highlight some aspects that should be included in the structure of CSR reports (Rego et al., 2007), namely: description of the company's strategy from a sustainable development point of view through the testimony of a board member; overall picture of the company's structure as well as its operations and markets; definition of the company's governance structure, inherent management policies and systems, including commitment to all stakeholders; elaboration of a content index based on the GRI report and measuring company impacts through economic, environmental and social performance indicators.

On the other hand, it is crucial that the company be concerned with socio-cultural norms, as they are determinants of behavior (Rego et al., 2007). When changes in societal norms occur, company behavior inevitably changes as well. From this perspective, it is relevant to consider that organizations that do not choose to adopt socially responsible behavior over time will have a strong chance of being excluded by customers and the community itself (Rego et al., 2007).

There are several factors behind the implementation of a CSR policy (CEC, 2001). First, there are recent concerns and expectations of citizens, consumers, public authorities and investors in a context of globalization and large-scale industrial change. In addition, social criteria have an increasing influence on individual or institutional investment decisions, whether as consumers or investors. There is also growing concern about the environmental damage caused by their economic activities. Lastly, mention should be made of the transparency resulting from business activities by the media and current information and communication technologies.

Lourenço and Schröder (2003) and Fernandes (2012) also present different motivations that they believe underlie CSR practices: improvement of the company's image and sales, due to the strengthening and loyalty to the brand or product; promoting labor and social rights in a context of globalization; increasing value of the company in society and in the market, as socially responsible companies attract more investors and are seen with greater credibility; reduction of taxation due to benefits granted to organizations that promote socially responsible actions; increased motivation and commitment of employees; pursuit of social profit; improving community living conditions, changing community attitude toward country problems, opportunities to take on a social intervention role and build a marketing strategy based on new business management trends.

On the other hand, if the company favors unethical and socially irresponsible behavior, it may face consequences such as: bad image and reduced sales of the company; removal of investors; negative publicity; customer complaints and loss of future consumers; payment of fines and compensation for violation of established laws and low productivity and employee demotivation.

According to the Green Paper, there is an intention by the European Commission to promote a European framework for CSR (CEC, 2001). This document highlights voluntary action by companies as a way to address environmental and social concerns raised by stakeholders, evident in the following quote:

> An increasing number of European companies are promoting their corporate social responsibility strategies as a response to a variety of social, environmental and economic pressures. They aim to send a signal to the various stakeholders with whom they interact: employees, shareholders, investors, consumers, public authorities and NGOs. In doing so, companies are investing in their future and they expect that the voluntary commitment they adopt will help to increase their profitability. (CEC, 2001, p. 3)

2.4 Social Responsibility Instruments and Tools: The SA8000 Standard

The SA8000 standard was created in 1997 by Social Accountability International, a nonprofit organization that operates in the field of ethics worldwide. According to the Social Accountability International Web site (2018a), SA8000 is based on labor provisions contained in the Universal Declaration of Human Rights and World Labor Organization (WTO) conventions. Social Accountability International provides organizations with various resources that support the maintenance and continuous improvement of their social performance, including capacity building, stakeholder engagement, cooperation between buyers and suppliers, as well as the creation of tools to ensure continuous improvement.

The SA8000 certification is conditional upon meeting the established requirements in order to fully respond to the interests and concerns of the organization's employees, subcontractors and even suppliers. The audit process is undertaken by independent entities. The main elements that constitute the norm relate to child labor, forced or compulsory labor, health and safety, freedom of association and the right to collective bargaining, discrimination, disciplinary practices, working hours, remuneration and management system (Social Accountability International, 2019)

Currently, Portugal has 39 certified companies operating in various sectors of activity, but this is still a small number compared to many countries (Social Accountability Accreditation Services, 2019). Although certification is a very costly, time-consuming and costly process, there are organizations that, regardless of their size, attach great importance to this certification. The certification lasts for three years and requires audits every six months during this period (Stigzelius & Mark-Herbert, 2009). The financial factor may be the major impediment for companies not to invest in certification (Leite & Rebelo, 2010).

In addition to the costs directly associated with certification, there are pre-certification activities that companies use, such as improving health and safety facilities, salary review, training and consulting. Internally, one of the biggest challenges for companies is raising wages and limiting overtime, knowing that consumers will not pay more for products even if the company complies with SA8000. Still at the company's internal level, Stigzelius and Mark-Herbert (2009) argue that another obstacle has to do with the difficulty in communicating the relevance of the standard to employees, which becomes a challenge for HRM.

On the other hand, several benefits from SA8000 certification are visible, including: improved management and performance of the value chain; building and reinforcing loyalty from employees, customers and other stakeholders and ensuring compliance with global standards, thereby reducing the risk of negligence, public exposure and possible legal action. Achieving certification also enables the company to prove its commitment to CSR and the ethical treatment of its workers in accordance with global standards. Certification also has a positive impact on building lasting ties, reducing turnover and absenteeism and increasing the ability to attract skilled labor (Stigzelius & Mark-Herbert, 2009).

Since the 2014 revision, the certification is based on the Social Fingerprint, a toolkit that enables companies to continuously measure and improve their social performance management system so that they can meet the requirements of the standard. There are ten categories associated with this tool (Social Accountability Internacional, 2018b):

1. Policies, Procedures & Records
2. Social Performance Team
3. Identification & Assessment of Risks
4. Monitoring
5. Internal Involvement & Communication
6. Complaint Management & Resolution
7. External Verification & Stakeholder Engagement
8. Corrective & Preventive Actions
9. Training & Capacity Building
10. Management of Suppliers & Contractors.

These categories are included in the self-assessment of the organization when applying to understand the level of maturity of its management system. Following the self-assessment, the accredited certification body conducts an assessment identifying the strengths and helping to improve them from an organizational management

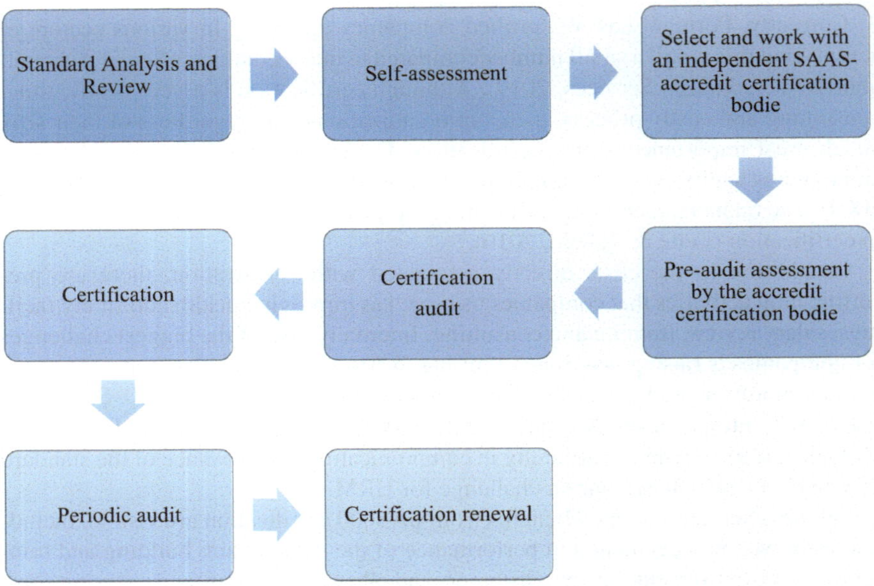

Fig. 4 Certification process. *Source* Adapted from Social Accountability International (2019)

perspective. Subsequently, the organization is subject to biennial audits (announced and unannounced) to ensure that management processes continue to comply with the requirements of the standard. Figure 4 shows, in an organized way, the inherent steps of the certification process:

3 Context and Research Method

This study adopts a qualitative methodology, inspired by the interpretative paradigm. The qualitative methodology aims to promote a full understanding of the phenomena, as well as the context in which they arise (Coutinho, 2013). Next, CSR is briefly contextualized in the Portuguese business reality.

3.1 The Context of CSR in Portugal

Although CSR takes into account social or environmental concerns in organizations' day-to-day procedures and relationships with stakeholders (customers, employees, suppliers, investors, etc.), it does not mean that the company should neglect the pursuit of profit, as it distinguishes from a third sector organization.

There is growing concern on the part of the European Commission on this issue, noted in the strategy presented in October 2011, through the document "A renewed EU strategy 2011–14 for Corporate Social Responsibility" (European Commission, 2011). This paper introduced a new definition of CSR, describing it as "the responsibility of enterprises for their impacts on society" (European Commission, 2011, p. 6). In order to fulfill their social responsibility effectively, companies need to be aware of social and environmental issues in order to create value for stakeholders and identify, avoid and mitigate their negative impacts (European Commission, 2011).

Regarding the Portuguese reality, in addition to the fact that companies have to comply with European guidelines such as the above, they are also governed by the Constitution of the Portuguese Republic, the Labor Code and the decree-laws and rules that cover the different types of CSR. The Constitution of the Portuguese Republic is the document that underlies the principles and organization of the Portuguese State and has established aspects related to CSR, whether internal or external for the company. Internally, aspects directly related to employees were recognized, such as hygiene, health and safety, dismissals, equality and discrimination, training, working conditions, salary, among others. Externally, the focus on the environment can be highlighted.

The Labor Code highlights some issues related to CSR, such as wages, special schemes (student-worker), hygiene, health and safety at work, redundancies, reconciliation between work and family life (paternity and maternity), equality and non-discrimination, among others (Labor Code, 2018) (Código do Trabalho, 2018). However, it is relevant to realize that the social aspect, particularly externally, is still the least explored from the legal point of view.

The Lisbon European Summit in March 2000 was a moment that contributed to the growing adoption of the concept of CSR by organizations. As evidence of this is that in 2002, Delta Cafés became the first Portuguese company to achieve SA8000 certification (Delta Cafés, 2014), followed by DHL which obtained certification in 2004. Since 2003, there has been a growing demand for certification from other companies, as organizations working on this topic have emerged due to the increasing number of events on the subject and the number of companies using consultants who provide such services (Branco & Rodrigues, 2007; Rego et al., 2007).

In the beginning of this century, CSR was at the time a concept still little known in Portugal (Pinto, 2004). According to Pinto (2004), 60% of companies admitted that they should pay more attention to this theme. In the same study, 70% of consumers stated that their purchasing decisions could be conditioned by the company's social behavior. While 28% would be willing to pay more for a socially and environmentally responsible product, 36% said they would not pay more (Pinto, 2004).

In 2008, Portugal first entered the KPMG Survey of Corporate Responsibility Reporting, which aimed to analyze the concern of organizations in reporting their CSR actions. Over the years, the country has been recognized for the quality of information and professionalism demonstrated by the companies (KPMG, 2017).

3.2 Research Method

Given the purpose of this work, it was first necessary to select companies certified and not certified by the SA8000 standard. From each of these categories, five companies were selected, all of them located in the North of Portugal. For certified companies, these could easily be identified through a list provided by the *Social Accountability Accreditation Services* (SAAS, 2019). In the case of non-certified companies, the choice was based on an electronic survey of a range of companies that worked CSR in their daily lives.

In each of the companies, the interviewee was the person responsible for the area of CSR, and there was no standard regarding the area of education, levels of education and position they held in the company. Table 1 presents the characterization of the interviewees. The interviews addressed questions about CSR, the definition of the concept, as well as the advantages and disadvantages associated with CSR. Questions were also elaborated on: practices promoted by the company (internally and externally), perception by stakeholders; how stakeholder expectations, interests and needs are met; how interest in CSR arose; assessment of returns on investments in CSR and future plans in this area. On the other hand, for certified companies,

Table 1 Interviewees characteristics

Interviewee	Age	Gender	Educational qualifications	Job role	Seniority in the company	Certified company/Not certified	Activity branch
1	43	Female	Law Degree	HR Director	3	Certified	Commerce of Electronic Material
2	49	Male	Degree in Management	Commercial Director	8	Certified	Food industry
3	42	Female	Degree in Business Administration	HR Coordinator	8	Not certified	Software development
4	45	Male	Degree in Biology	Head of Social Responsibility Standard	18	Certified	Waste management
5	68	Male	Degree in Anthropology	HR Director	6	Not certified	Construction
6	52	Female	High School	Assistant Director	18	Certified	Textile industry
7	37	Female	Law Degree	HR Director	3	Not certified	Weighing systems
8	31	Male	Degree in Education	Marketing Director	4	Not certified	Automobile trade
9	40	Male	Master in Humanities	HR Director	4	Certified	Textile industry
10	37	Female	Degree in Social Communication	People & Culture Department Director	2	Not certified	Metal industry

there was an interest in understanding in more detail the importance of SA8000 certification, how the certification process took place and the implications it had for stakeholders. In the case of non-certified companies, questions were defined that aimed to understand if there was already interest on the part of the company to obtain certification and why, and what advantages do they consider to have with certification.

The interviews lasted an average of 40 min and always took place at the premises of the various companies. After conducting the interviews, they were transcribed. Dimensions of analysis were defined, divided into subcategories, which directly related to the script questions. Table 2 summarizes the dimensions.

Subsequently, the data content analysis was performed. To facilitate the entire analysis process, the MAXQDA software was used. The last stage of this investigation reflects the result of content analysis, using the technique used—the interview—establishing a relationship with the literature, in order to enable a response to the problem.

Table 2 Summary of analysis dimensions

Dimension	Aspects addressed by dimension
Company contextualization	Company characteristics (number of employees, size) Strategy and organizational identity Existence of certain organizational documents and knowledge of these by stakeholders
Implementation of CSR policies and practices	Importance of CSR for the company Personal representation of aspects related to the concept Identification of CSR practices streamlined by the company and how they are perceived by stakeholders Internal and external communication Evaluation of current practices and returns from investing in CSR Future perspectives on CSR
SA8000 certification	Personal representation of the importance of the existence of SA8000 standard Importance of the company's Social Responsibility certification (certified) Certification process (certified) Stakeholder involvement in certification; Advantages of obtaining certification (not certified) Desire to obtain certification (not certified)

4 Findings

In this section, the analysis of the findings is presented, with a focus on the main dimensions as highlighted by the interviewees and according with the objectives of this study.

4.1 Companies Contextualization

4.1.1 Companies Characteristics

With regard to this study, the companies involved are mainly SMEs (10 and 250 employees). The fact that the companies are SME is relevant since they represent the majority of the Portuguese businesses, and Rego et al. (2007) argue that SMEs play an active role in the communities where they operate, supporting social causes and contributing to their social and economic development. Therefore, CSR can be available to all types of organizations, regardless of their nature, size or line of business.

Only three of the companies interviewed distinguish themselves as large companies, as they have more than 250 employees. It is noteworthy that this is a universe of companies in different states of development and with different longevities (between 10 and 60 years of activity).

The level of certification in the group of certified 'companies includes three medium-sized companies and two large companies. In this study, a small company, two medium-sized companies and only one large company represent the uncertified companies.

4.1.2 Strategy and Organizational Identity

CSR demands a constant commitment and coherence from the organization, reflected in its strategy and organizational identity, as these are the elements that differentiate CSR and philanthropy (Correia, 2013). This concern can be evidenced more clearly in the values that govern the organization, by alluding to ethical and socially responsible behaviors. Interviewees stated various values that accurately portray their ethical and socially responsible stance, ranging from integrity, loyalty, rigor and transparency in business to commitment and value creation for stakeholders. Respondents 8 and 9 reinforce the strong need to give back to the community, which is the basis of their success and should belong to the company's DNA. The success of the company is explained by the greater retention and sense of belonging of the employees. In the words of interviewee 9: "We have differentiating practices and this is what makes us have people here for 40 years (…) our average age is 45 years, 75% of people have been with this company for over 11 years, so this is the return we can get from the practices we implement."

Interviewee 7 mentioned that her values are based on a motto that guides her daily life, in which everyone sees themselves as a family, which translates into the existence of affective bonds, whether with the internal or external community. Interviewee 5 emphasizes the issue of labor conflicts and is proud that, in recent years, he has not resorted to labor litigation. This does not mean that there are no conflicts, it happens that these situations are easily resolved, based on the dialogue and values defended by the company, according to interviewee 5.

Regarding the mission, there is a discourse based on the most profitable aspect of the company, although in some interviews, there is reference to aspects that go beyond the economic and legal responsibilities of the companies (Carroll, 1979, 2016). In the words of interviewee 10, "both in mission, vision and values we focus people a lot, because for us this is the essential." Interviewee 10 justifies this growing concern about people was the reason for the creation of the People & Culture Department, which aims to make exactly what the company prioritizes. According to interviewee 1, the mission is clearly influenced by the company's concern with the entire internal environment, that is, customer satisfaction is only guaranteed if the employee is satisfied and motivated: "a collaborator who is well will show that and customer will be much more satisfied." For this reason, the company continually strives to ensure full reconciliation between work and family life. Interviewee 6 corroborates this argument, stressing the importance of teamwork as a way to achieve better performance.

4.1.3 Organizational Documents

While ethical behavior and socially responsible behavior appear to be unquestionable, it is crucial that companies explicitly demonstrate this in their organizational documents. To this end, it is critical that companies develop a code of ethics and conduct that accurately portrays their fundamental objectives, values and responsibilities to their stakeholders (CEC, 2001; European Commission, 2011).

Since the code of ethics is a legally binding document, there is obvious unanimity in the responses. This document is presented to stakeholders as a way to raise awareness of the importance of commitment. In the case of employees, the document may be diluted in the employment contract or may be provided at the time of its integration. However, one of the interviewed companies also opted to include some ethical principles in the Collective Bargaining Agreement, that is, an agreement between the company and one or several trade unions in the industry in question. There are also companies that focus on internal training that covers these issues in order to provide a general understanding of the organization's principles of ethical conduct.

From the perspective of Balonas (2014), it is very important for companies to draw up a social report that reflects the environmental and social impact of companies, as well as the quality of their relationship with their stakeholders. Currently companies already incorporate some of the social balance indicators in the so-called Single

Report, a document that is required of all companies. However, some companies go beyond this simple document and specifically develop a social report that serves as a mechanism for planning, coordinating and rationalizing available resources (Rego et al., 2007). Interviewee 5 admits that the company he represents prepares a social report, which is displayed, but considers that "[...] people are sometimes not very interested." (2007) when he states that "[the social report] presents things that should be considered, it helps us to understand some difficulties we have" (interviewee 5). By way of example, interviewee 6 points out that, even if there is no specific document, at the moment, some of the social balance indicators are being analyzed for the implementation of a company training program in different areas, which allow personal and professional development.

Rego et al. (2007) also argue that, currently, there has been a growing curiosity among stakeholders to access information related to the company, which reflects its behavior in a global way. According to the interviewees, this interest is fundamentally on the part of their customers and suppliers, given the business relationships they seek to establish.

4.2 Implementation of CSR Policies and Practices

4.2.1 Importance of CSR for the Company

It is critical that CSR be intrinsic and genuine, in line with company strategy. Therefore, it will be incorrect to implement CSR policies and practices only to follow the trend of other companies or to cover up certain company bad practices (Rodrigues et al., 2009). Interviewee 5 presents a view that reflects the literature when he states that "Social Responsibility cannot be the reverse of remorse (...) it must be exactly the social commitment between peers, between those who own the means of production and those who provide the service."

The remaining respondents report that the main objective associated with CSR is to "create value" (interviewee 3) through "a set of activities that go beyond what is legal" (interviewee 9), "in order to create quality, for both the internal audience and the surrounding community" (interviewee 3). Interviewee 2 also adds that "companies do not live in isolation, they live together with society, so they have to be aware of the difficulties and needs that society lives in."

In general, all respondents referred to the role of management in maintaining ethical and socially responsible behavior, and this concern is similarly taken into account in certified and non-certified companies. Interviewee 4 reinforces the need for management involvement by saying that "things need to be done, but if management is not involved it will not be possible to do it, because resources are needed, people must be available (...), this link is needed with the administration." In the case of the company represented by interviewee 10, only one of the directors initially supported the implementation of CSR practices, but once they began to see the results of this investment, the other members of management quickly came to support and get involved in the process.

4.2.2 Corporate CSR Policies and Practices

CSR has gained increasing relevance in organizations, contributing to better business orientation and, consequently, to the well-being of society (CEC, 2001; European Commission, 2011), which is true of the companies that participated in this study. Thus, the existence of two dimensions in which CSR practices are inserted is distinguished: the internal dimension and the external dimension. In some companies, there is no balance between these two dimensions, which means that the practices implemented are mostly internal or external. Interviewee 1 emphasizes the need to get employees to understand the needs of the community, when he says: "we try to take our employees abroad and see that Social Responsibility is much more than what we live inside."

The internal dimension is distinguished by its focus on the organization itself, namely in the areas of HRM, health and safety at work, adaptation to change and management of environmental impact and natural resources (CEC, 2001). Attracting and retaining qualified talent is one of the biggest challenges facing organizations today, so it is imperative that some practices emerge that seek to meet their needs in order to improve their performance. All companies have recognized some practices that follow this view, ranging from designing spaces and leisure activities for employees (respondents 1, 4, 5, 7, 9), to awarding prizes or other benefits such as life and health insurance (respondents 1, 3, 5, 7, 8, 9). Some respondents also stress the existence of practices of reconciling family and professional life (respondents 1, 3) or making food and meals available to employees (interviewees 4, 6, 7, 8, 10). From these statements, it appears that there is a constant effort on the part of companies to maintain a good working environment, offering good conditions and meeting the needs of their employees. In the words of the interviewees:

> … we care about the well-being of employees, that they have the best working conditions (…) we make sure that people are satisfied with working conditions and that they are motivated. (Interviewee E6)

> … We try to meet the major changes in the labor market, particularly in terms of family, professional / personal conditions and motivation. (Interviewee 3)

> … We take great care of what our employees need, have and do. (Interviewee 4)

It is important that companies truly face employees in their individuality, creating conditions for them to assume their family and social responsibilities. Interviewee 1 mentions this as one of the top priorities for the company: "management attaches great importance to reconciling work and personal life, because a worker who is concerned about what he has left behind will not have the profitability we want (…) The employee knows that he has support here when he needs it."

From the point of view of employees' skills development, companies also refer to the existence of training, whether it is directed to soft skills (transversal skills) or hard skills (technical skills). In the opinion of interviewee 5, "vocational training is also a matter of social responsibility (…) that makes workers grow, regardless of their category, whether they are qualified or not qualified." In turn, the interviewee 1 admits that training is a development mechanism for young people who integrate the company through curricular and professional internships and, after that period, have great possibilities of being able to enter into an employment contract with this entity.

Interviewee 6 refers to the importance given to participatory management by the company that represents: "We have suggestion and idea boxes in which all employees are free to come up with their ideas, anyone can have an idea, and we try to implement those ideas." However, she adds that she would like this practice to have a greater adherence by the employees.

In some interviews, environmental practices were highlighted, which aim to reduce the environmental impact of industrial processes and the distribution of goods and services (CEC, 2001; European Commission, 2011). For example, interviewee 9 states that "we have photovoltaic panels, we replace our car fleet with electric cars, we take care of waste management properly (…) we want to leave a footprint on our planet in order to make some difference." Interviewee 6 refers to the organic garden that the company owns, which in addition to fostering employees' knowledge of seasonal products, also serves to provide meals or distribution to employees of the organization.

The companies interviewed also have a large external participation, either with their business partners or with local communities. Almost unanimously, the companies mentioned regular support for institutions, initiatives, social projects or families in need through monetary or goods donations. However, interviewees 7 and 10 admit the need for rigorous selection of requests arriving at the company and subsequent monitoring of their actual destination: "Social Responsibility is not just about giving a donation (…) the intention is that that donation will serve something bigger in the future, it is like the example of fish and fishing rod (…) should give a fishing rod and teach how to fishing" (interviewee 7). Volunteering is also a highly valued activity by some companies through initiatives such as building houses for families with poor economic conditions (interviewees 1 and 8), supporting people in their homes (interviewee 2), food collection campaigns (interviewee 3), blood donations (interviewees 7 and 3) and fundraising activities for institutions (interviewee 2). It is also noteworthy that, according to interviewee 4, there is in the organization a "volunteer association that works both internally and externally."

Interviewee 8 admits that volunteering can be a major contribution to creating links between employees. In this sense, interviewee 2 also recalls that "the company itself often gives up a few hours of work for employees to engage in volunteering." So, they are both giving up; the employee is giving a little of himself and the company because lets the employee do it.

From an ethical point of view, interviewee 6 considers the implementation of the "supplier kit" relevant, where the company seeks to disclose its policy, mission, principles and its code of conduct, in order to "keep suppliers informed of what is done and how important it is to work with people with the same goals." The same interviewee also adds: "We also seek to instill in our suppliers this spirit as well, that we want to be responsible for a better service, for a better life for all" (Interviewee 6).

With regard to encouraging entrepreneurship, the companies that stand out at this level are those represented by interviewees 2 and 10. According to the first, activities are promoted in partnership with some schools, in order to promote entrepreneurship in young people. Interviewee 10 highlights the creation of a manufacturing unit in the complex of one of its companies in Mozambique, where users of a local institution develop their business: "we set up, bought equipment (…) at the beginning we help with material and they are now sustainable and all profit is for them… it is also a way of supporting the institution."

Two of the interviewees also include in their portfolio some initiatives that are more focused on employees' children. Concerning interviewee 9, the company prioritizes the applications of employees' sons in its recruitment processes, considering that this contributes to the "image of a solid company." On the other hand, interviewee 6 emphasizes the initiative of exchanging textbooks for the children of employees, or in case this reuse is not possible, the company donates the books to some institution."

However, there is still little adherence to certain practices, as mobilization of all employees is sometimes difficult (interviewee 3). The main reasons for this poor adherence may be related to the fact that, in certain activities, employees do not identify with their purpose or because they do not consider they have the skills to participate (interviewee 6). Interviewee 2 highlights a weak participation in only one of the volunteer projects that the company promotes, related to home social support. However, there are other practices where adherence is huge. Interviewee 10 somewhat contradicts the previous testimonies, when she states that there is a great involvement, especially in the collection of goods, by all stakeholders: "last year we had a furniture collection action for a family affected by the fires (…). People really worked hard." Still, it appears that different stakeholders view CSR differently, and according to interviewees, those who value and care most are customers and workers.

When asked about their CSR performance, respondents revealed that while they are moving in the right direction, there is still a lot of work that needs to be done, especially because CSR is an ongoing process. Regarding the physical spaces of the company, the interviewee 10 states that they will soon start building a new production unit, where there will be a greater concern at the environmental level, in order to make the building more sustainable and environmentally friendly.

The truth is that corporate investment in CSR policies and practices provides a great set of advantages (as presented by Lourenço & Schröder, 2003; Mendonça & Gonçalves, 2004; Fernandes, 2012) and Lievens & Slaughter, 2016), and that were clearly recognized by all respondents. The main advantages pointed out by the interviewees are social recognition and company branding. In addition to the previous advantages, they add the differentiation of these companies to their customers, when they already value this issue (interviewees 6 and 10) or even the fact that these companies may be influencing their stakeholders (interviewee 10). Internally, there are advantages associated mainly with the motivation and satisfaction of employees, which will be reflected positively in their performance (interviewees 4 and 9). For respondent 9, this performance should always be recognized, as the company values a culture of meritocracy, where the best are differentiated without questioning the principle of equal rights.

4.2.3 Evaluation of Current Practices and Returns from Investing in CSR

After listing the advantages and disadvantages pointed out by the interviewees, it was proposed to make an overall balance regarding the returns resulting from the investment in CSR practices. Therefore, it must be remembered that we are dealing with entities where the goal is profit maximization. However, in these organizations, CSR practices emerge in the form of investment, which is constantly subjected to a cost–benefit analysis. Interviewees 3 and 5 consider this return to be difficult to measure specifically in some in-house initiatives but believe it is warranted and generally reflected in the organization:

> We do not quantify, we are sure that the return is guaranteed, especially by social peace, and this is priceless. (Interviewee 5)
>
> It is difficult to measure (…) those who participate in volunteer actions participate motivated by principles such as kindness, generosity (…) I believe that people will feel better doing that (…) internally, it is easy to understand that the return is higher than what is invested (…) it has to do with internal motivation, an employee who has health insurance paid by the company is at least more motivated or when has a problem gets more rested than if he had not. (Interviewee 3)

Lourenço and Schröder (2003), Fernandes (2012) and Lievens and Slaughter (2016) argue that companies that implement CSR practices benefit from an improvement in their image and a higher appreciation in society, increasing their credibility:

> The impact this has on the customer is very large, and it is already starting to appear (…) we are already evaluated in Social Responsibility (Interviewee 10).
>
> We have a daily customer demand (…) and I think it has to do with it (Social Responsibility practices) (…) people are looking for, seeing and liking our profile (…) we have customers that we have had for many years, and This is very important (Interviewee E6).

On the other hand, the same authors also mention that, by investing in CSR practices, companies may aim to obtain social profit but are also focused on providing better living conditions to the community or assuming a role of social intervention. Interviewee 10's testimony corroborates this argument using an example of a collection of goods that they organized where, despite the costs involved for transportation, they managed to provide better living conditions for the inhabitants of Mozambique.

The literature portrays a strong relationship between social marketing and CSR, which is manifested in most interviews. From an overview, companies seek through marketing to enhance their awareness, improve their branding and reputation, which will enable them to achieve greater external recognition and retention of talent, regardless of whether they are certified or uncertified. Some respondents pointed to some tools such as corporate TV (interviewee 4 and 6) and the internal newsletter (interviewee 4).

4.3 SA8000 Certification

SA8000 certification requires organizations to comply with certain requirements that relate to child and forced labor, health and safety, freedom of association and the right to collective bargaining, non-discrimination, disciplinary practices, working hours, remuneration and management system.

From the perspective of the interviewed representatives of the certified companies, this authentication is a reflection of the daily work performed in the companies. For interviewee 4, the certification "allows us to have a photograph of the situation (…) in terms of organization (…) we get a much clearer photograph, that is, we say but it is different if someone else says, both to us (…) as after externally." Interviewee 6 refers to certification as the "added value" given to the organization, which gives it a competitive advantage over its competitors. It also admits that, once certified, companies have an additional responsibility, namely for being committed to the requirements of the standard.

For non-certified companies, the importance of the standard is equally recognized, as stated by respondents 3 and 5:

> I think the standard is extremely important (…) it will help define processes, procedures within the scope of Social Responsibility. (Interviewee 3)

> This external self-verification gives some assurance, and when you ask for certification, it is not so much to use in the office of the administrator, but as required by all employees… it cannot be rhetorical… the more certifications, the more control we have in compliance with objectives. (Interviewee 5)

When asked about their willingness to obtain certification, some respondents revealed that this is one of their goals: "We have, in fact, been thinking about this lately. There is no decision yet, whether or not we are moving forward, but it is a situation that is under consideration (…) no doubt that in the future we will move forward… I cannot say that it will be this year or next" (interviewee 7); "We have some certification processes underway, and this is also part of the portfolio" (interviewee 5). Interviewee 3 confirms that there is a need to link the company to certain processes; however, for reasons that relate to strategic management decisions, the certification process has not yet been completed.

Subsequently, this group of respondents was asked about the advantages they considered to derive from SA8000 certification. Following the examples cited by Stigzelius and Mark-Herbert (2009), interviewees indicated that certification can bring advantages such as commitment to CSR (interviewee 5), market differentiation (interviewee 10) and the creation of a standard of procedures and practices (interviewees 3 and 8).

4.3.1 Certification Process

In the context of certified companies, it is important to understand how the certification process went. This is a process that involves different stages up to the time of certification and is therefore considered by some interviewees as "laborious" (interviewees 1 and 4) and "very bureaucratic" (interviewees 1 and 6), although not a slow process.

Great stakeholder involvement is required to facilitate the implementation of some procedures. All interviewees reported that there is a lot of involvement from various stakeholders, from employees to customers and suppliers. As indicated in one of the previous topics, management should always be primarily involved, not only in implementing CSR practices, but also in the decision to obtain certification. From there, there will be a greater willingness of workers and various departments to be involved in this process.

According to interviewee 4, at an early stage, they sought to make an internal diagnosis, in partnership with an external company, which aimed to assess the degree of compliance with the assumptions of the standard. It adds that the involvement with suppliers and subcontractors, although constant, was intensified at a later stage, due to the imposition of the standard to audit these partners. In interviewee own words: "There are a number of things that they have to do and have to be involved in… in the implementation itself, with the exception of suppliers control, the rule is more internal, so there was no great need to be involved initially." However, some respondents assume that there is still some resistance, particularly from suppliers or customers, because they are subject to audits. So, it can be said that this is really the biggest challenge for certified companies: "Our *Achilles Heel* is precisely that, our suppliers are the hardest part" (interviewee 6); "When we start getting into suppliers, things get complicated… we have to try to get them to do the same things we do, which is very difficult (…) is a very sensitive area and we have to continue to work on it in the future" (interviewee 4).

On the other hand, interviewees 6 and 9 confess that there is a great deal of openness from some companies with whom they do business because they share the same goals. According to interviewee 6, "they do not all react the same way, because many do not actually follow this evolution or are not certified and do not want to invest in this area (…) but most, fortunately, are meeting our goals, they are very receptive and supportive." In turn, interviewee 9 considers that "it is an easy process, usually the companies we have worked with are open to this way of thinking."

In short, the involvement of all stakeholders in the certification process is essential, and they should first be made aware of the requirements of the standard and then should be taken into account in its implementation.

4.3.2 Implications of Certification for Companies

Certification is a very exhaustive and demanding process, which implies above all a great commitment from the organization. However, the certification results are essentially related to the differentiation and external recognition. According to interviewee 9, "the norm is a matter of showing the world that we are socially responsible (…) this is the most important part of having a document that certifies us, because if we did not have it, we would not stop doing what we do (…) but there are clients who only work with us because we are socially responsible."

According to the interviewees, the certification process did not require major changes in the social responsibility practices developed by the company, as these practices were already part of the organization's daily life. In the words of interviewee 1, "only those who are involved in the most bureaucratic parts can see the dynamics inherent in certification (…) employees do not realize much because it is already part of their routine."

Workers can see certification as giving them some security and comfort. This safety and comfort arise, in particular, from the obligation imposed by the 2014 revision of the standard regarding the existence of a social performance team, which includes a workers' representative, in addition to the administration, managers, union representatives and employees involved in the certification (interviewee 4). In this sense, interviewee 6 also highlights the holding of periodic meetings between workers and their representative, in order to know their concerns.

However, respondents admitted that certification allows them to diagnose opportunities for improvement or correct certain behaviors. According to interviewees, it "allows us to quickly identify anything that is not going well (…) or what can be improved and act quickly" (interviewee 4); "Even though we are audited we have zero non-conformities in SA8000 (…) but there may be one or another situation that could be improved" (interviewee 9).

From the perspective of interviewee 1, the standard confines, to a certain extent, the freedom existing in the organization, due to the need to strictly comply with certain requirements. Interviewee 1 argues that it "it is a barrier, because we have more freedom without the standard (…) given most companies I think the standard is essential, given the reality of our company, maybe it was not necessary."

In summary, it can be seen that CSR practices have always been something that was part of the daily lives of these companies, thus not requiring a major change when obtaining certification. Thus, there is no evidence to confirm a clear influence of certification on CSR practices.

5 Final Remarks

In Portugal, CSR has gained increasing visibility over time. Companies are increasingly interested in implementing policies and practices with goals that go beyond profit, aimed at stakeholders such as workers, suppliers and the local community (Lopes & António, 2016). However, this process has been gradually developed in companies because it is still difficult to measure its impact in the social context.

Still, there is a need for organizations to recognize their social responsibility, not only for their economic interests, but also for the needs and interests of different stakeholders, viewing CSR as a management paradigm rather than a fad. There are companies that adopt CSR spontaneously due to organizational culture, ethical orientation of owners and managers, or strategic choice. In fact, what is often heard about CSR is that "it is not always what it seems," because certain similar organizational behaviors may mirror completely different strategies and goals, where some are characterized as being genuinely ethical and others as purely instrumental (Rego et al., 2007). However, the implementation of policies and practices, as well as certification, is only possible if all stakeholders embrace the principles, since only in this way will they be actively involved. In addition, companies should be mindful of the adoption of good and appropriate HRM practices that respect the individuality and personal and professional dignity of their employees.

While SA8000 certification gives the company a greater competitive advantage, it is noteworthy that this is a very expensive process, which indicates that this is not an easily accessible process for companies. As it is a norm that is very directed to the internal aspect of the organization, it might be expected that there would be a greater number of internal dimension practices in certified companies, which is not the case, since there are also several practices directed to the local community. This may imply that these companies really do not seek to do the minimum required to obtain certification but rather seek to invest in the external side as well, as this is an intrinsic concern of the company. On the other hand, it was found that there are no major discrepancies regarding the implemented practices between certified and non-certified companies, and some of the practices are even similar. Thus, it is essential that companies, certified or not, find innovative strategies to meet the expectations and needs of their stakeholders.

In the future investigations, it would be appropriate to use a larger sample that is representative in order to obtain general conclusions on the subject. On the other hand, it would be equally interesting to involve a greater diversity of branches of activity, so that it would be possible to make clear comparisons of the reality of various sectors of activity.

Acknowledgements Delfina Gomes acknowledges that this study was conducted at the Research Center in Political Science (UID/CPO/0758/2019), University of Minho/University of Évora, and was supported by the Portuguese Foundation for Science and Technology and the Portuguese Ministry of Education and Science through national funds.

References

Azevedo, J. B., Ende, M. V., & Wittmann, M. L. (2016). Responsabilidade Social e a Imagem Corporativa: O Caso de uma Empresa de Marca Global. *Revista Eletrónica de Estratégia e Negócios, 9*(1). Disponível em: http://portaldeperiodicos.unisul.br/index.php/EeN/article/view/3142.

Balonas, S. (2014). Olhar o Público Interno: O Fator Crítico nas Estratégias de Responsabilidade Social. Relatório de um debate Centro de Estudos de Comunicação e Sociedade. Em T. Ruão, R. Freitas, P. Ribeiro, & P. Salgado (Edits.). Braga: Centro de Estudos de Comunicação e Sociedade.

Blowfied, M., & Murray, A. (2008). *Corporate responsibility: A critical introduction*. Oxford: Oxford University Press.

Branco, M. C., & Rodrigues, L. L. (2007). Positioning stakeholder theory within the debate on Corporate Social Responsibility. *Electronic Journal of Business Ethics and Organization Studies, 12*(1), 5–15. Disponível em: http://ejbo.jyu.fi/pdf/ejbo_vol12_no1_pages_5-15.

Carroll, A. B. (1979). A three dimensional conceptual model of corporate social performance. *Academy of Management Review*, 497–505.

Carroll, A. B. (2016). Carroll's pyramid of CSR: Taking another look. *International Journal of Corporate Social Responsibility, 1*(3), 1–8.

Código do Trabalho. (2018). Coimbra: Almedina.

Comissão Europeia. (2002). *Regional clusters in Europe*. Luxemburgo: Observatory of European SMEs.

Commission of the European Communities (CEC). (2001). Green paper: Promoting a European framework for Corporate Social Responsibility (CEC). Available at: http://www.europarl.europa.eu/meetdocs/committees/deve/20020122/com(2001)366_en.pdf. Accessed December 09, 2019.

Correia, A. S. (2013). *A Responsabilidade Social e as PME: As Práticas de RSE das Microempresas*. Dissertação de Mestrado, Instituto Politécnico de Lisboa, Instituto Superior de Contabilidade e Administração de Lisboa, Lisboa.

Costa, M. A. (2005). Fazer o Bem Compensa? Uma Reflexão sobre a Responsabilidade Social Empresarial. *Revista Crítica de Ciências Sociais, 73*, 67–89.

Coutinho, C. P. (2013). *Metodologia de Investigação em Ciências Sociais e Humanas: Teoria e Prática* (2ª ed.). Coimbra: Almedina.

Delta Cafés. (2014). *Relatório de Sustentabilidade: Rostos de uma Marca*. Disponível em: http://www.deltacafes.pt/DeltaFiles/content/201602/fkkqakdc.5mh_df55f26e_contentfile.pdf?_ga=2.93847444.887861192.1550829912-1700029022.1544693899.

Duarte, A. P., Mouro, C., & Neves, J. G. (2010). Corporate Social Responsibility: Mapping its social meaning. *Management Research: The Journal of the Iberoamerican Academy of Management, 8*(2), 101–122.

Elkington, J. (2004). Enter the triple bottom line. Em J. R. Adrian Henriques, *The Triple Bottom Line: Does It All Add Up?* (pp. 1–16). London: Routledge.

European Commission. (2011). *A renewed EU strategy 2011–14 for Corporate Social Responsibility*. Available at: https://www.europarl.europa.eu/meetdocs/2009_2014/documents/com/com_com(2011)0681_/com_com(2011)0681_en.pdf. Accessed December 13, 2019.

Fernandes, J. (2012). *A Responsabilidade Social das Empresas – Uma Alavanca para a Sustentabilidade? Um Estudo de Caso: O Grupo Nestlé e as Plantações de Cacau na Costa do Marfim*. Tese de Mestrado em História, Relações Internacionais e Cooperação, Faculdade de Letras da Universidade do Porto.

Gardner, H. (2006). *The development and education of the mind: The selected works of Howard Gardner*. London and New York: Routledge.

Ethos Instituto. (2017). *Indicadores Ethos para os Negócios Sustentáveis e Responsáveis*. São Paulo: Instituto Ethos.

Jorge, F., & Silva, M. L. (2011). Responsabilidade Social e Desenvolvimento Local Sustentável. *Conferência Responsabilidade Social no Desenvolvimento das Comunidades Locais e Regionais*. CLDS de Évora - Contrato Local de Desenvolvimento Social de Évora.

Kotler, P., & Lee, N. (2005). *Corporate Social Responsibility—Doing the most good for your company and your cause*. New Jersey: Wiley.

KPMG. (2017). *The road ahead: The KPMG survey of corporate responsibility reporting*. KPMG.

Leal, A. S., Caetano, J., Brandão, N. G., Duarte, S. E., & Gouveia, T. R. (2011). *Responsabilidade Social em Portugal*. Lisboa: bnomics.

Leandro, A., & Rebelo, T. (2011). A Responsabilidade Social das Empresas: Incursão ao Conceito e Suas Relações com a Cultura Organizacional. *Exedra, Número Especial*, 11–39. Disponível em: http://www.exedrajournal.com/docs/s-CO/01-11-40.pdf.

Leite, C., & Rebelo, T. M. (2010). Explorando, Caracterizando e Promovendo a Responsabilidade Social das Empresas em Portugal. *Actas do VII Simpósio Nacional de Investigação em Psicologia* (pp. 2209–2225). Universidade do Minho.

Lievens, F., & Slaughter, J. E. (2016). Employer image and employer branding: What we know and what we need to know. *The Annual Review of Organizational Psychology and Organizational Behavior, 3*, 407–440. Disponível em: https://biblio.ugent.be/publication/8100868/file/8100876.pdf.

Lopes, M. M., & António, N. J. (2016). Responsabilidade social empresarial em Portugal: Do Mito à Realidade. *Internacional Business and Economics Review, 7*, 110–138.

Lourenço, A. G., & Schröder, D. D. (2003). Vale Investir em Responsabilidade Social Empresarial? Stakeholders, Ganhos e Perdas. *Responsabilidade Social das Empresas: A Contribuição das Universidades, 2,* 77–119.

Mascarenhas, M. P., & Costa, C. D. (2011). Responsabilidade Social e Ambiental das Empresas. Uma Perspectiva Sociológica. *Latitude, 7*(2), 141–167. Disponível em: http://www.seer.ufal.br/index.php/latitude/article/view/1013/pdf.

Mendonça, J. R., & Gonçalves, J. C. (2004). Responsabilidade Social nas Empresas: Questão de Imagem ou Essência? *Organizações & Sociedade, 11*(29), 115–130.

OECD. (2004). *Principles of corporate governance.* Paris: OECD. Disponível em: http://www.oecd.org/corporate/ca/corporategovernanceprinciples/31557724.pdf.

Perrini, F. (2006). SMEs and CSR theory: Evidence and implications from an Italian perspective. *Journal of Business Ethics, 67,* 305–316.

Pinto, G. R. (2004). Responsabilidade Social das Empresas – Estado da Arte em Portugal - 2004. Lisboa: Centro de Formação Profissional para o Comércio e Afins (CECOA).

Rego, A., Cunha, M. P., Costa, N. G., Gonçalves, H., & Cabral-Cardoso, C. (2007). *Gestão Ética e Socialmente Responsável.* Lisboa: Editora RH.

Rodrigues, J., Seabra, F., & Ramalho, J. (2009). Contributos para uma Clarificação do Conceito. Responsabilidade Social das Organizações. *Cadernos Sociedade e Trabalho, 11,* 99–105.

Schwartz, M. S., & Carroll, A. B. (2003). Corporate Social Responsibility: A three-domain approach. *Business Ethics Quarterly, 13*(4), 503–530.

Social Accountability Accreditation Services. (2019). SA8000 Certified Organisations. Acedido a 9 de abril de 2019, em http://www.saasaccreditation.org/certfacilitieslist.

Social Accountability Internacional. (2018a). SA8000:2014. Acedido a a 26 de novembro de 2018, em http://www.sa-intl.org/index.cfm?fuseaction=Page.ViewPage&pageId=1711.

Social Accountability Internacional. (2018b). SA8000® Standard. Acedido a 26 de novembro de 2018, em http://www.sa-intl.org/index.cfm?fuseaction=Page.ViewPage&pageId=1929.

Social Accountability International. (2019). SA8000 Standard. Available at: http://www.sa-intl.org/index.cfm?fuseaction=Page.ViewPage&PageID=1689. Accessed December 11, 2019.

Stigzelius, I., & Mark-Herbert, C. (2009). Tailoring corporate responsibility to suppliers: Managing SA8000 in Indian garment manufacturing. *Scandinavian Journal of Management, 25,* 46–56.

Wood, J. D. (1991). Corporate social performance revised. *Academy of Management Review, 16*(4), 691–718.

Nexus Between Sustainable Development and CSR—An Empirical Study on Indian Nationalized Banks

Sudin Bag, Nilanjan Ray, and Atanu Manna

Abstract The nature of the business houses has got a new shape in the present arena. The globalization effects on the business drastically and emphasis has been given to the growth of the business with sustainability. The modern business has changed its motives and focused on the development with sustainability. The concept of development with sustainability has become more significant after the companies bill 2013 came into existence. According to this bill, it is mandated contribution of 2% of the average profit for 3 years. It is obvious that business houses are exploiting and using the recourses from the society, and it is expected to work for the betterment of the society, planet and stakeholders. The present study focuses on the efforts of Indian banks in their sustainable development. Further, the aim of this study is to analyse the relationship between CSR and profitability of Indian banks. The study is based on the secondary data which have been collected from the annual reports of the Reserve Bank of India as well as the report of selected Indian banks. Twenty nationalized banks have taken into account in this study. The collected data have been tabulated to represent the percentage and average as well. The correlation analysis has been performed to find out the relationship between CSR and profitability of Indian banks. The results show that the CSR activities have improved over the years and banks are concentrating the development with sustainability. The regression analysis shows that there is a significant relationship between the CSR and sustainability in terms of profitability of the Indian banks.

Keywords CSR · Sustainability · Profitability · Development · Indian banks

---

S. Bag
Department of Business Administration, Vidyasagar University, Midnapore, India
e-mail: sudinbag1@gmail.com

N. Ray (✉)
School of Management, Adamas University, Kolkata, India
e-mail: drnilanjanray.mgmt@gmail.com

A. Manna
Research Scholar, Centre for Environmental Studies, Vidyasagar University, Midnapore, India
e-mail: atanumanna329@gmail.com

C. Machado and J. P. Davim (eds.), *Circular Economy and Engineering*,
Management and Industrial Engineering,
https://doi.org/10.1007/978-3-030-43044-3_4

1 Introduction

In recent decades, corporate social responsibility and sustainable development achieved more popularity in business management and economic science. Sustainable development is the adjustment process between human and environment for meeting the present needs and without compromising the capability of future generations to fulfil their own needs. While corporate social responsibility is a flexible business model that supports a company to be socially responsible to itself, stakeholders and the public. These concepts have nothing any visible boundary to understand the ultimate definition of corporate social responsibility. The relationship between sustainable development and corporate social responsibility measurement is a very complex assignment. Baumgartner (2004) proposed a generic concept for business management with key dimensions and implications for strategic management, structure of organization and culture. Dawkins and Lewis (2003) mentioned that corporate responsibility is synonym of CSR, where responsibility includes employee welfare, ethics, environment and community commitment.

The sustainable development especially in macro level consists by three pillars under the triple bottom line approach. These are social, environmental (ecological) and economic (financial). These three pillar concepts are called in the corporate sector as corporate sustainability. Among three pillars social aspect is called CSR (Baumgartner and Ebner 2006). Moreover, Fergus and Rowney (2005) present semantic frameworks which clearly explain the terms and logical meanings of sustainability and its three dimensions as it is properly understood today. Welford (2005) stated that CSR is a business concept and he presents twenty elements such as internal and external aspects, accountability and citizenship which should be fulfilled by CSR-companies. Whereas Bazin and Ballet (2004), Zambon and Del Bello (2005) argued that without recognizing the needs of the stakeholders, not a single company can act in a social responsible way. Therefore, the concepts of sustainability are the natural evolution on the basis of stakeholder theory. Gauthier (2005), Knox et al. (2005) and Moir (2001) also identified CSR as an ethical basis which should be satisfying the needs of the stakeholders. CSR is business models that add in sustainable development by providing social, environmental and economic advantages for all stakeholders. The broadness of CSR in recent decades leads it towards the sustainable development (Financial Times).

CSR activities are considered as a way of sustainable profit and increase brand image among the stakeholders in the banking sector (Kaur, 2016; Moharana, 2013). It is fact that many Indian banks are doing CSR from their inception. At the same time, a few Indian banks initiate CSR for achieving competitive advantage and improve their brand image or reputation in the business perspective. According to Asian Sustainability Ranking (ASR) on social enterprise, CSR India is ranked fourth among the subcontinent countries which are drawing the importance towards CSR among the Indian banking institution. Through the wide range of financial product and services, the banking sector can expand the CSR activity for the sustainable development in India.

Banking sector mainly contributes in rural development like community welfare, women and children, health, etc (Kaur, 2016). The economic performance of the banking sector greatly contributes and accelerated in the rural transformation (Bhat & Bhatt, 2018). Many banks taking CSR as an optional basis and do not disclose any activities under CSR scheme. It is also true fact that a good number of banks are strong in CSR philosophy (Kaur, 2016).

2 Review of Literature

The word CSR came in the reality in the late 1960s and early 1970s. Arlow and Gannon (1982) expressed in their study that the business organization should involve in resolving the social problems even if it is not associated with any financial benefits.

The Indian banking sector has taken an initiative to make sure that banks are doing CSR in an organized and efficient manner. Poverty eradication, health and medical care, financial inclusion, education, rural area development, self-employment training, financial literacy training, etc. are some of the important areas where banks are concentrated (Dhingra and Mittal 2014).

According to Narwal (2007), the Indian banks have an objective viewpoint about corporate social responsibility (CSR) activities. Indian banks are concentrating mainly on education, health, balanced growth, environmental marketing and customer satisfaction as their core CSR goals. The Indian banking industry is found to be adopting an integrated approach by combining CSR with the ultimate customer satisfaction. Irrespective of location, the ultimate focus of CSR activities undertaken by banks is found to be similar.

Das (2012) suggested that corporate social responsibility (CSR) developed slowly in India though it was started a long time ago. Corporate social responsibility has assumed greater importance in the corporate world, including the banking sector. There is a visible significant trend in the financial sector by promoting eco-friendly and socially responsible lending and investment practices. The Govt. of India is pursuing the matter relating to corporate social responsibility and also drafted guidelines for corporate social responsibility practices time to time. The study also concluded that all the financial institutions including banking sector in India are directly engaged in social banking and development banking approach. Indian banks are mainly focused on education, rural development, women empowerment, financial support to weaker sections and helping the physically challenged.

According to Eliza (2013), the Indian banks are doing well in the corporate social responsibility areas, but still there is need of more emphasis on corporate social responsibility. There are some banks, which are not even meeting the regulatory requirements. The public sector banks have an overall highest contribution in corporate social responsibility activities. Private sector banks and foreign banks are still lagging behind in this area.

Divya and Aggarwal (2013) suggested that the corporate social responsibility has emerged as a benchmark for measuring the corporate excellence in the context of national and international banking business practices. The banking industry has a significant change with corporate social responsibility concept execution in India.

3 Objectives of the Study

From the literature review, it is assured that many researchers have lighten the issue of CSR in Indian banking sectors but no one has identified the ultimate result of CSR in Indian banking sector. Based on the above literature and research gap, the objectives of this paper are as follows:

1. To show the CSR concentration zone of Indian bank.
2. To measure the relationship between CSR and SD with respect to Indian bank.
3. To identify the effectiveness of CSR on SD in India from the viewpoint of Indian bank.
4. To measure the progress rate of CSR in Indian banks.

4 Research Methodology

In order to recognize the CSR pattern of a specific area, a study of CSR combination is imperious. The CSR amalgamation on the one hand provides an idea about CSR typology, CSR economics (financial), CSR environmental and CSR sociological aspects of the area, and on the other hand, it provides a perception into the CSR practices and initiative of CSR, which are quite significant for the maintenance of societal issues. The concept of CSR combination of factors is a scientific method to study the prevailing relationships of factors in association with each other in corporate social responsibility of business environment. Its demarcation is not a culmination in itself, but only a tool on the way to a better understanding of the corporate social responsibility situations.

The present study is based on secondary data which have been collected from the annual report of the twenty (20) nationalized banks in India and report of Reserve Bank of India (RBI) has also taken into account in this study. The SPSS-22, Excel and Statgraphics 18 softwares have been used by the researcher to analyse the data and design the concentration zone of CSR activities in Indian banks.

A CSR concentration zone or combination of aspects of the banking sector was calculated through the real percentage of the spending amount in three pillars (social, economic and environment) against the theoretical standard value. For standard measurement, the theoretical curve was calculated as follows:

- Mono aspect hypothetical value = 100%

- Double aspect hypothetical value = 50%
- Triple aspect hypothetical value = 33.33%.

For determination in which pillar the banks are mostly concentrated, the standard deviation method is used. Here, relative value is more important than absolute value. So here square root is not extracted and the formula is as follows:

$$\sigma = \sqrt{\frac{d^2}{n}}$$

where d = (Hypothetical value percentage − Actual value percentage)

n = no. of aspect. (Like Mono aspect = 1, Dual aspect = 2, Triple Aspect = 3)

To establish the relationship between CSR and SD in the banking sector, Pearson correlation coefficient method is used. In this case, all the possible factors of CSR and SD are exploited for determining the CSR and SD value. The formula is used as follows:

$$r = \frac{n(\sum xy) - (\sum x)(\sum y)}{\sqrt{[n \sum x^2 - (\sum x)]\left[n \sum y^2 - (\sum y)^2\right]}}$$

To establish the effectiveness of CSR on SD, the formula is as follows:

Effectiveness of CSR on SD

$$\text{Ecsr} = \frac{\text{No. of factors associated with CSR in Banking Sector}}{\text{Total no. of possible SD factor}}$$

Effectiveness value:

The value of effectiveness always consists of 0 to 1

where 1 = mostly effective,

0.5 = moderately effective,

0 = non-effective.

CSR progress rate:

$$\text{CSR Progress Rate} = \frac{\text{CSRpresent} - \text{CSRpast}}{\text{CSRpast}} * 100$$

5 Analysis and Interpretation

In this section, the researcher focused on the concentrated zone of CSRs in Indian nationalized banks. The Indian banks are spending a remarkable amount as the social responsibility with the aim of sustainability development. The CSR's activities of banking sectors are broadly categorized into three main pillars, namely social, economical and environmental perspectives. The total amount spent by the Indian banks under the umbrella of CSR in the major three pillars is represented in Table 1.

From Table 1, the researcher has calculated concentration zone or combination of aspects, i.e., which of the above-mentioned pillars are mostly concentrated by the Indian banks as CSR activities. The concentration zones have been calculated by mono aspect, dual aspect and triple aspect. Chronological order of percentage value helps to calculate the aspects. The mono aspect implies that the Indian banks are only concentrating any one of the three aspects (i.e., social/economical/environmental), while dual aspect means any two of the pillars are focused by the Indian banks as

Table 1 Amount spent (Rs. in crore) by Indian banks under CSR activities

Sl. No.	Bank Name	Social		Economical		Environmental	
		2018	2019	2018	2019	2018	2019
1	Allahabad Bank	0.58	1.13	7.69	5.25	0	0.06
2	Andhra Bank	1.84	1.15	1.1	0.56	0.25	0
3	Bank of Baroda	0	5.95	0	5.95	0	0
4	Bank of India	3.47	0.44	0.98	0.65	1.97	1.09
5	Bank of Maharashtra	0.29	0	0	0	0	0
6	Canara Bank	30.35	25.62	2.65	3.03	0.14	0.28
7	Central Bank of India	0	0	0	0	0	0
8	Corporation Bank of India	0.19	0.43	1.86	2.8	0	0
9	Dena Bank	0	0	2.43	2.55	0	0
10	Indian Bank	1.04	3.56	1.44	1.42	0.11	0.71
11	Indian Overseas Bank	0	0	0	0	0	0
12	Oriental Bank of India	0.93	2.81	0	0	0.3	1.4
13	Punjab National Bank	1.71	1.6	2.59	25.4	2.73	1.61
14	Punjab & Sind Bank	0	0	0.24	0	0	0
15	Syndicate Bank	0.8	0.64	0	0.38	0.65	0.38
16	State Bank of India	101.33	101.82	3.96	5.3	3.94	3.7
17	UCO Bank	8.85	8.77	11.73	0.66	0.25	0
18	Union Bank	7.27	4	0	0.67	0	0
19	United Bank of India	2.27	1.2	0	1.2	0.57	0.79
20	Vijaya Bank	3.98	2.29	0	0.56	0.98	1.13

Source Annual Reports

CSR activities and triple aspect indicate the Indian banks are concentrated all the three pillars as the CSR activities for sustainability. An illustration of Vijaya Bank 2019's CSR activity is depicted below as an example.

For Mono aspect:

$$= \sqrt{\frac{(100 - 57.54)^2}{1}}$$
$$= 42.46$$

Dual aspect:

$$= \frac{(50 - 57.54) + (50 - 28.39)}{2}$$
$$= 16.18$$

Triple aspect:

$$= \sqrt{\frac{(33.33 - 57.54)^2 + (33.33 - 28.39)^2 + (33.33 - 14.07)^2}{3}}$$
$$= 18.08$$

Table 2 shows which combination best suited for the Indian banking sector. So among three aspects lowest value represents the combination of aspects.

From Table 2, it is clear that in 2018, three banks, namely Bank of Baroda, Central Bank of India and Indian Overseas Bank did not spent any amount under the CSR activity. However, the Bank of Maharashtra, Canara Bank, United bank of India, Corporation bank, Allahabad bank, Vijaya bank, State bank of India, Oriental Bank of India and Union banks focused on mono social aspect for CSR activities, whereas Dena Bank, Punjab & Sind Bank also concentrated on mono economical aspect as CSR activities. The dual aspect of CSR activities has been taken by the Bank of India, Andhra bank, Indian bank, Syndicate bank and UCO Bank. Punjab National Bank is concentrated on the triple aspect of CSR activities. For the better understanding of CSR concentration zone of Indian bank, the research has developed a *Ternary Diagram* which is represented below.

The three pillars of CSR activities (2018), namely social, economical and environmental, are represented in Fig. 1 and the zone of concentration of Indian banks are also depicted by the various symbols corresponding to the banks. The range of these three pillars runs from zero (0) to one hundred (100). Zero indicates no concentration of the banks towards the particular CSR activity, and hundred represents the full concentration. Here, Punjab National bank, Andhra Bank and Bank of India belong to the triple aspect of CSR activities for the sustainable development in India.

Table 3 represents the amount spent by Indian banks under the various CSR activities in the year of 2019. It is very clear that the concentrated area of Indian banks

Table 2 Percentage of amount spent in 2018 under three pillars and their spending aspects

Sl No	Bank name	Percentage of amount spent			Combination		
		Social	Economic	Environment	Mono	Dual	Triple
1	Allahabad Bank	7.01	92.99	0.00	7.01	42.99	
2	Andhra Bank	57.68	34.48	7.84	42.32	12.24	20.36
3	Bank of Boroda	0	0	0	0	0	0
4	Bank of India	54.05	15.26	30.69	45.95	13.95	15.95
5	Bank of Maharastra	100.0 0	0.00	0.00	100.00		
6	Canara Bank	91.58	7.99	0.42	8.42	41.79	
7	Central Bank of India	0	0	0	0	0	0
8	Corporation Bank of India	9.27	90.73	0.00	9.27	40.73	
9	Dena Bank	0.00	100.00	0.00	100.00		
10	Indian Bank	40.24	55.64	4.12	44.36	7.97	21.59
11	Indian Overseas Bank	0	0	0	0	0	0
12	Oriental Bank of India	75.61	0.00	24.39	24.39	25.61	
13	Punjab National Bank	24.30	36.90	38.80	61.2	12.18	6.43
14	Punjab & Sindh Bank	0.00	100.00	0.00	100.00		
15	Syndicate Bank	55.17	0.00	44.83	44.83	5.17	
16	State Bank Of India	92.77	3.63	3.61	7.23	44.61	42.03
17	Uco Bank	42.50	56.30	1.20	43.7	6.93	23.41
18	Union Bank	100.0	0.00	0.00	100.00		
19	United Bank Of India	79.93	0.00	20.07	20.07	29.93	
20	Vijaya Bank	80.24	0.00	19.76	19.76	30.24	

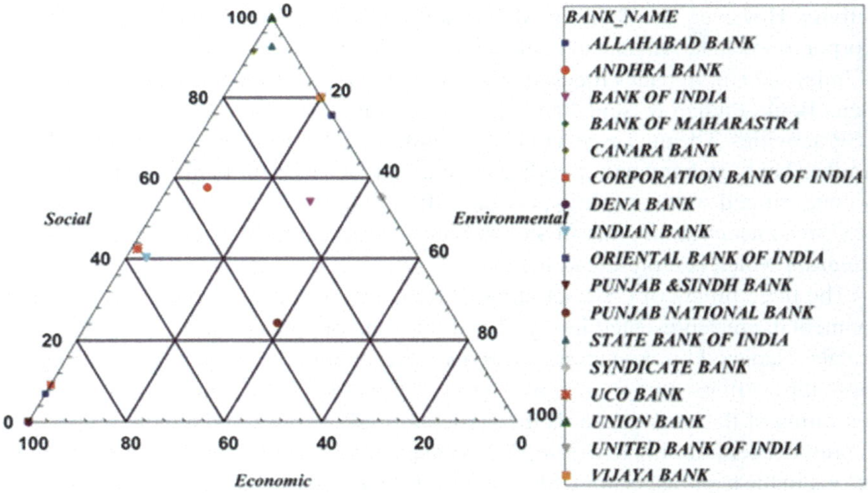

Fig. 1 CSR activity of Indian banks in 2018

Table 3 Percentage of amount spent in 2019 under three pillars and their spending aspects

Sl No	Bank name	Percentage of amount spent			Combination		
		Social	Economic	Environment	Mono	Dual	Triple
1	Allahabad Bank	17.59	81.47	0.94	18.26	31.94	
2	Andhra Bank	67.25	32.75	0.00	32.75	17.25	
3	Bank of Boroda	50.00	50.00	0.00	50.00	0.00	
4	Bank of India	20.00	30.00	50.00	50.00	14.14	12.47
5	Bank of Maharastra	0	0	0	0	0	0
6	Canara Bank	88.56	10.47	0.97	11.44	39.05	
7	Central Bank of India	0	0	0	0	0	0
8	Corporation Bank of India	13.31	86.69	0.00	13.31	36.69	
9	Dena Bank	0.00	100.00	0.00	100.0		
10	Indian Bank	62.60	24.93	12.47	37.40	19.84	21.31
11	Indian Overseas Bank	0	0	0	0	0	0
12	Oriental Bank of India	66.75	0.00	33.25	33.25	16.75	
13	Punjab National Bank	5.59	88.78	5.63	11.22	41.66	39.20
14	Punjab & Sindh Bank	0	0	0	0	0	0
15	Syndicate Bank	45.71	27.14	27.14	54.29	16.44	8.75
16	State Bank of India	91.88	4.78	3.34	8.12	43.58	41.40
17	Uco Bank	93.00	7.00	0.00	7.00	43.00	
18	Union Bank	85.65	14.35	0.00	14.35	35.65	
19	United Bank of India	37.62	37.62	24.76	62.38	12.38	6.05
20	Vijaya Bank	57.54	14.07	28.39	42.46	16.18	18.08

has been in comparison with the year 2018. A good number of Indian banks focus on dual as well triple aspects of CSR for sustainable development. It is interesting that three banks (earlier it was only one) concentrated on the triple aspect of CSR activities and another five banks concentrated on dual aspect CSR activities for the sustainability. Therefore, the five banks are really looking for the sustainable development. On the other side, it is also found that only four banks are not focused on the any CSR activities. The following ternary diagram represents the concentration zone of Indian banks for the year 2019.

Figure 2 represents the CSR activities of Indian banks in the 2019. From the ternary diagram, it is clear that five banks, namely Bank of India, Punjab National Bank, Syndicate Bank, United Bank of India and Vijaya Bank, are concentrated on the triple aspect of CSR activities for sustainability in India.

From the ternary Figs. 1 and 2, we can classify 17 Indian banks into 7 categories according to the nature of CSR concentration and group into then as social, economical, environmental, socio-economic, socio-environmental, eco-environmental and socio-eco-environmental. The classification of Indian banks is depicted in Table 4.

Category of the ternary Fig. 3.

From Fig. 3, it is clear that the concentration zone of Indian national banks is significantly different from each other. UCO Bank, Union Bank, Bank of Baroda and Punjab National Banks have given emphasis on the social issues with respect

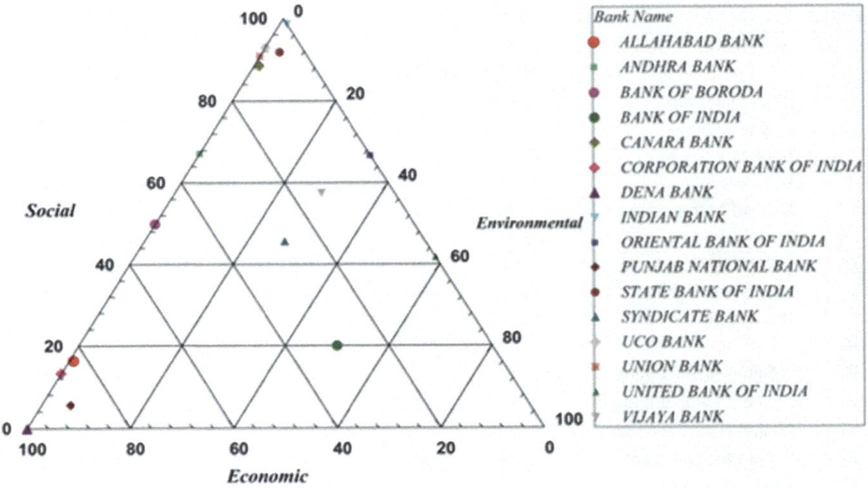

Fig. 2 CSR activity of Indian banks in 2019

Table 4 CSR concentration zones of Indian banks

Sl No	Bank Name	2018				2019			
		Class			Category	Class			Category
		M	D	T		M	D	T	
1	Allahabad Bank		√		Socio-Economy		√		Socio-Economy
2	Andhra Bank			√	Socio-Eco-Env		√		
3	Bank of Boroda	x	x	X			√		Socio-Economy
4	Bank of India			√	Socio-Eco-Env			√	Socio-Eco-Env
5	Bank of Maharastra	√			Social	x	x	x	
6	Canara Bank		√		SOcio-Economy		√		Socio-Economy
7	Central Bank of India	x	x	X		x	x	x	
8	Corporation Bank of India		√		Socio-Env		√		Socio-Env
9	Dena Bank	√			Economical	√			Economical
10	Indian Bank		√		Socio-Economy				Socio-Economy
11	Indian Overseas Bank	x	x	X		x	x	x	
12	Oriental Bank of India		√		Socio-Env		√		Socio-Env
13	Punjab National Bank			√	Socio-Eco-Env			√	Socio-Eco-Env
14	Punjab & Sindh Bank	√			Economical	x	x	x	
15	Syndicate Bank		√		Socio-Env			√	Socio-Eco-Env
16	State Bank of India	√			Social		√		Socio-Economy
17	Uco Bank		√		Socio-Economy		√		Socio-Economy
18	Union Bank	√			Social		√		Socio-Economy
19	United Bank of India		√		Socio-Env			√	Socio-Eco-Env
20	Vijaya Bank		√		Socio-Env			√	Socio-Eco-Env

M-Mono aspect, D-Dual aspect, T-Triple aspect

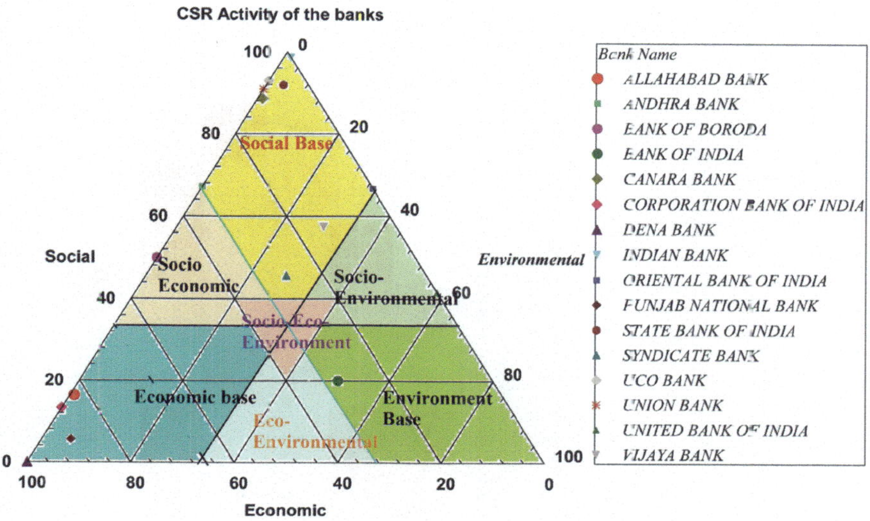

Fig. 3 CSR activity of the banks

to corporate social responsibility. The environmental as well as economical concentration is not so sound with respect to CSR activities of the above-mentioned four banks in India.

On the other side, Allahabad Bank and Canara Bank are focusing on the economic perspective as an activity of CSR for the sustainable development. At the same time, United Bank of India, Syndicate bank, Bank of India, Vijaya Bank and Punjab National Bank have given emphasis on the three aspects with respect to corporate social responsibility (CSR) for the sustainability in India.

The correlation analysis has been performing to access the relationship between the CSR and SD. We have taken ROE, Price Earnings Ratio (PER), Market Cap-Sales Ratio (PSR), Profit Before Interest and Taxes (PBIDT), Earnings Per Share (EPS), Net Interest Margin (NIM), Profit Before Tax (PBT), Profit After Tax (PAT), Net Profit Ratio (NPR) and RONW as indicator of sustainable development of Indian banks. From Fig. 4, it is found that there is positive and significant relationship among the CSR, PAT, PBT, PBIDT, PSR, NPR in the 2017. The relationship is significant at 1 and 5% level. But it is also a fact that there is no significant relationship between CSR and ROE, PER and RONW. Thus, it is obvious that CSR activities help in developing financial health which is the indicator of sustainable development.

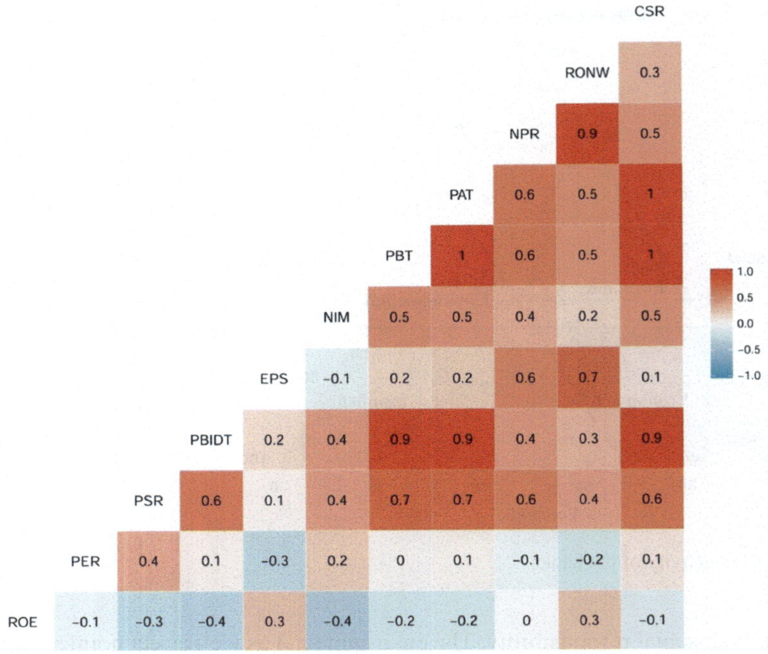

Fig. 4 Correlation analysis to access the relationship between CSR and SD in the year 2017. Note: CSR corporate social responsibility, RONW return on net worth, NPR net profit margin, PAT profit after tax, PBT profit before tax, NIM net interest margin, EPS earning per share, PBIDT profit before interest and taxes, PSR market cap-sales ratio, PER price earnings ratio and ROE return on equity

From Figs. 5 and 6, it is clear that CSR has a significant relationship with PBT, PAT, PBIDT and PSR at 1% level of significance. But there is no relationship among the CSR, NPR, ROE, PER and RONW in the years 2018 and 2019, respectively.

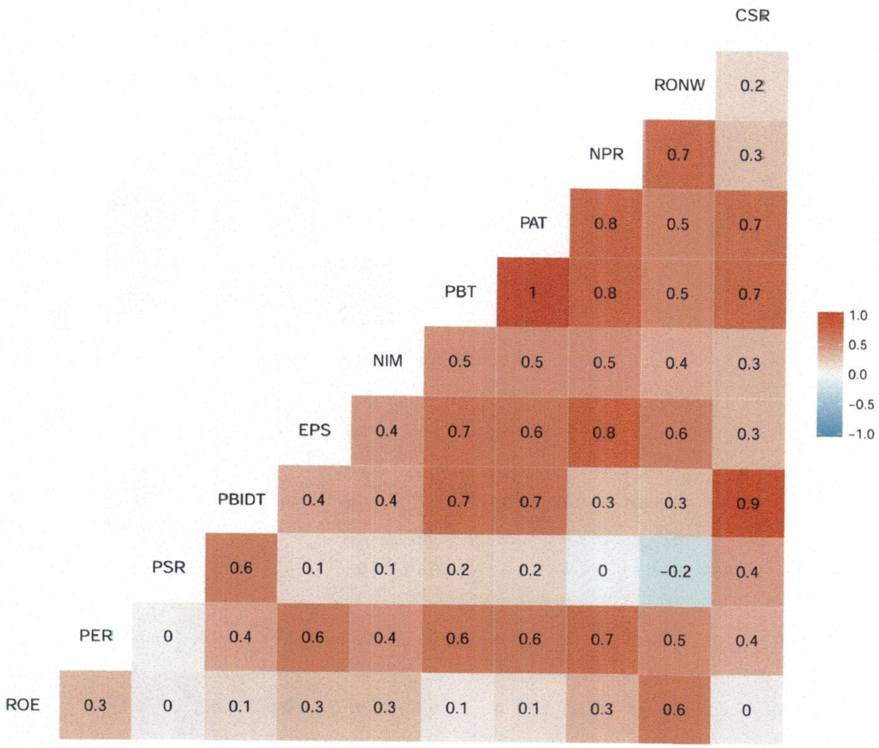

Fig. 5 Correlation analysis to access the relationship between CSR and SD in the year 2018

6 Calculation of Effectiveness of CSR on SD

Katara and Arora (2014) on their paper identified 19 major areas where the banking sector mainly concentrates for development of the society. Those 19 major areas were identified for calculating the ACSR. The major areas were educational support, poverty eradication, rural development, farmer's club, village knowledge centre, joint liability groups for promotion of SHGs, adoption of the girl child, microfinancing, environmental issue, rural development, women's empowerment, vocational training, community welfare, physically challenged, sustainability, corporate volunteering, social investment, health and child welfare.

90 indicators are taken as a major area of SD, which is used for SD calculation according to the IISD, Canada.

No. of factors was calculated on the basis of CSR activities in no. of different sectors doing by the Indian banking sector.

Illustration: Vijaya Bank 2018–19 no. of CSR and effectiveness of CSR on SD

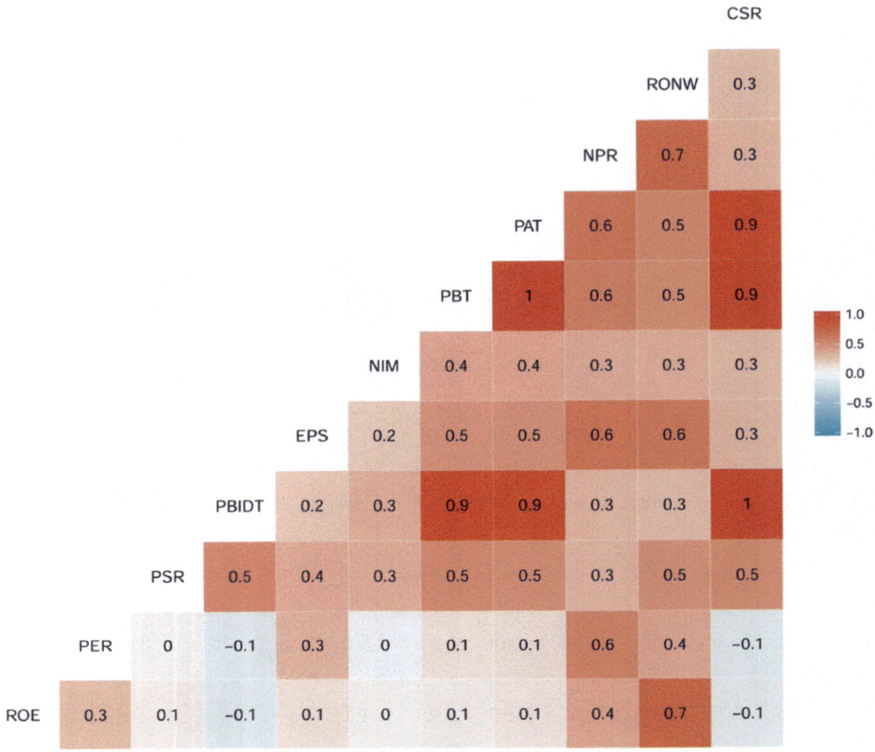

Fig. 6 Correlation analysis to access the relationship between CSR and SD in the year 2019

$$\text{Ecsr} = \frac{12}{90} = 0.133$$

Table 5 shows the effectiveness of CSR on SD. When the value of effectiveness less than 0.5, implies that the effectiveness of CSR on sustainable development is not so sound. In the perspective of sustainable development, the CSR activities mainly focus on the societal development rather than the economical as well as environmental development.

Progress Rate Measurement

On the basis of methodology, the progress rate is calculated (Table 6).

Example: Vijaya banks 2017–18 and 2018–19 data are used for illustration of CSR progress rate

$$\text{CSR progress Rate} = \frac{(4.96 - 2.73)}{2.73} * 100 = 81.68$$

Table 5 Effectiveness of CSR on sustainability development

Bank name	No. of factors 2018	No. of factors 2019	Effectiveness 2018	Effectiveness 2019
Allahabad Bank	4	11	0.044	0.122
Andhra Bank	6	7	0.067	0.078
Bank of Baroda	4	5	0.044	0.056
Bank of India	3	7	0.033	0.078
Bank of Maharashtra	0	0	0.000	0.000
Canara Bank	7	10	0.078	0.111
Central Bank of India	0	0	0.000	0.000
Corporation Bank of India	5	10	0.056	0.111
Dena Bank	3	3	0.033	0.033
Indian Bank	8	10	0.089	0.111
Indian Overseas Bank	0	0	0.000	0.000
Oriental Bank of India	9	9	0.100	0.100
Punjab National Bank	10	11	0.111	0.122
Punjab & Sind Bank	1	2	0.011	0.022
Syndicate Bank	11	13	0.122	0.144
State Bank of India	11	14	0.122	0.156
UCO Bank	10	9	0.111	0.100
Union Bank	9	12	0.100	0.133
United Bank of India	2	4	0.022	0.044
Vijaya Bank	12	7	0.133	0.078

From Table 6, researchers can say that the Punjab National Banks is doing maximum progress in CSR activities during 2018–2019 and the growth rate is almost 952%. The progress rate of Oriental Bank of India and Indian bank is also remarkable which is 242 and 115%, respectively, in the year 2019. Apart from that Corporate bank of India, Dena Bank, State Bank of India and United Bank of India are also doing positive CSR activities for sustainable development in India. It can also be seen from Table 6 that the progress rate of the banks is not significant and it is negative, i.e., the banks have reduced the CSR spending amount in the year 2019 in comparison with the previous year.

Table 6 CSR progress rate of Indian banks

Bank Name	2017	2018	2019	Progress rate 2018	Progress rate 2019
Allahabad Bank	10.99	8.5	6.45	−22.66	−24.12
Andhra Bank	1.28	3.19	1.71	149.22	−46.39
Bank of Baroda	19.46	0	11.91	−100.00	0
Bank of India	16.95	6.42	2.18	−62.12	−66.04
Bank of Maharashtra	0	0.29	0	0	−100.00
Canara Bank	5.68	33.18	28.98	484.15	−12.66
Central Bank of India	6.6	0	0	−100.00	0
Corporation Bank of India	4.76	2.05	3.24	−56.93	58.05
Dena Bank	8.3	2.43	2.55	−70.72	4.94
Indian Bank	2.96	2.65	5.72	−10.44	115.88
Indian Overseas Bank	0.38	0	0	−100.00	0
Oriental Bank of India	0	1.23	4.21	0	242.28
Punjab National Bank	6.76	2.72	28.61	−59.76	951.84
Punjab & Sind Bank	0	0.24	0.15	0	−37.50
Syndicate Bank	12.93	1.45	1.4	−88.79	−3.45
State Bank of India	114.92	109.82	112.96	−4.44	2.86
UCO Bank	4.23	20.85	10.41	392.91	−50.07
Union Bank	1.22	7.27	4.67	495.90	−35.76
United Bank of India	2.13	2.77	3.19	30.05	15.16
Vijaya Bank	2.73	4.96	3.98	81.68	−19.76

7 Conclusion

At the nutshell, the researcher can conclude that out of the twenty Indian nationalized banks seventy banks are concentrating on the corporate social responsibility (CSR) activity for the sustainable development. Among the three pillars, namely social, economical and environmental, the stronger pillar is society as the highest concentration zone of CSR activity of Indian banking sector. The other two pillars are also getting importance slowly as the CSR concentration zone. To increase the financial sustainability, Indian banks are merely concentrated on the educational training, women empowerment, micro financing, rural development, vocational training, community welfare, corporate volunteering, social investment, health and child welfare. Corporate social responsibility of the banking sector in increasing dramatically and CSR activities positively leads to the sustainable development.

References

Arlow, P., & Gannon, M. J. (1982). Social responsiveness, corporate structure, and economic performance. *The Academy of Management Review, 7*(2), 235–241.

Baumgartner, R. J. (2004). Sustainable business management: Conceptual framework and application. In: Y. A. Hosni, R. Smith & T. Khalil (Eds.), *IAMOT (International Conference on Management of Technology)*. Washington.

Baumgartner, R. J., & Ebner, D. (2006). The relationship between sustainable development and corporate social responsibility. In: *Corporate Responsibility Research Conference*. Dublin. http://www.crrconference.org.

Bazin, D., & Ballet, J. (2004). Corporate social responsibility: the natural environment as a stakeholder? *International Journal of Sustainable Development, 7*(1), 59–75.

Bhat, A., & Bhatt, M. (2018). Gender equality for sustainable development: A distant dream in India? *Asia Pacific Journal of Research, 1*(28), 24–29.

Das, S. K. (2012). Practices of corporate social responsibility (CSR) in banking sector in India: An assessment. *Research Journal of Economics*.

Dawkins, J., & Lewis, S. (2003). CSR in stakeholder expectations: And their implication for company strategy. *Journal of Business Ethics, 44*(2–3), 185–193.

Dhingra, D., & Mittal, R. (2014). CSR practices in Indian banking sector. *Global Journal of Finance and Management, 6*(9), 853–862.

Divya, M., & Aggarwal, M. (2013). CSR - a strategy for sustainable business success: Evidence from Indian companies. *Journal of Commerce and Business Studies*, Delhi School of Economics, University of Delhi.

Eliza, S. (2013, February). Corporate social responsibility: An analysis of Indian commercial banks. *AIMA Journal of Management & Research, 7*(1/4). ISSN 0974 – 497

Fergus, A. H. T., & Rowney, J. I. A. (2005). Sustainable development: Lost meaning and opportunity? *Journal of Business Ethics, 60*(1), 17–27.

Gauthier, C. (2005). Measuring corporate social and environmental performance: The extended life-cycle assessment. *Journal of Business Ethics, 59*(1–2), 199–206.

Katara, S., & Arora, L. (2014). Emerging trends in CSR in Indian banks. *International Journal of Multidisciplinary Consortium, 1*(3), 136–143.

Kaur, S. (2016). A study on Corporate Social Responsibility (CSR) in Indian banking sector. *International Journal of Current Research, 8*(11), 42604–42606.

Knox, S., et al. (2005). Corporate social responsibility: Exploring stakeholder relationships and programme reporting across leading FTSE companies. *Journal of Business Ethics, 61*(1), 7–28.

Moharana, S. (2013). Corporate social responsibility: A study of selected public sector banks in India. *IOSR Journal of Business and Management (IOSR-JBM), 15*(4), 1–9. e-ISSN: 2278-487X, p-ISSN: 2319-7668.

Moir, L. (2001). What do we mean by corporate social responsibility? *Corporate Governance, 1*(2), 16–22.

Narwal, M. (2007). CSR initiatives of Indian banking industry. *Social Responsibility Journal, 3*(4), 49–60.

Welford, R. (2005, Spring). Corporate social responsibility in Europe, North America and Asia. 2004 survey results. *The Journal of Corporate Citizenship, 17*, pp. 33–52.

Zambon, S., & Del Bello, A. (2005). Towards a stakeholder responsible approach: The constructive role of reporting. *Corporate Governance, 5*(2), 130–141.

How to Look at Organizations and Human Resource Management in the Economy of the Future?

Diana Fernandes and Carolina Feliciana Machado

Abstract The present work seeks to provide a theoretical reflection, based on a bibliographical review of the main theoretical and empirical contributions on the influence of social, economic, financial, political, environmental, and other external forces on human resources management (HRM) approaches. According to the logic of Vickers (Hum Resour Plan 28(1):26–32, 2005) and Jabbour and Santos (Int J Hum Resour Manag 19(12):2133–2154, 2008), in presenting what we conceive as the present and future of HRM, based on the concept of "Social Pollution" (Pfeffer in Acad Manag Perspect 24(1):34–45, 2010), we refer to sustainability as an imperative. We intend to launch the discussion on how human resources management policies and practices can be guided in the sustainability vector, in order to contribute to the improvement of the social and environmental performance of organizations, that is, adding value of social relevance. We suggest, based on the contributions of Dryzek (The politics of the earth: environmental discourses. Oxford University Press, Oxford, 2005) and following the work of Ferrão (Que Economia Queremos? Fundação Francisco Manuel dos Santos, Lisboa, 2014), that currently two main discourses of change coexist: the green growth economy and the welfare economy. To guide the thinking, we propose a theoretical model. The key implication of the model is that an organization's ability to generate revenue from resources will depend primarily on its effectiveness in managing the context (internal and external). Therefore, the drawing and implementation of such conscious HRM models will allow to see organizations as spaces for joint promotion of opportunities in which each employee is considered as a co-creator of solutions and not as a simple executor of tasks and functions (Coutinho and Pereira in Urban Stud 39(13):2395–2411, 2010). Another implication is that future research on sustainable competitive advantage must focus not only on the attributes of tangible resources but also on how tangible and intangible resources are developed, managed and disseminated. Efforts to identify sources of resource capital and institutional capital between competitors can shed additional light on the management of both types of capital in order to foster sustainable competitive advantage, underscoring the relevance of longitudinal studies on the development and deployment of resources.

D. Fernandes · C. F. Machado (✉)
School of Economics and Management, University of Minho, Braga, Portugal
e-mail: carolina@eeg.uminho.pt

C. Machado and J. P. Davim (eds.), *Circular Economy and Engineering*,
Management and Industrial Engineering,
https://doi.org/10.1007/978-3-030-43044-3_5

Keywords HRM · Circular economy · Organizations · Economy · Future

1 Introduction

Organizations consist of dynamic institutions, made up of people. So, and in order to achieve a sustainable organization it will be necessary to embrace financial, economic, social, cultural, political, and environmental impacts of its operations, which implies, first of all, a focus on the analysis of its human resources management (HRM) model.

The effects of management practices on social and physical well-being have until recently tended to be largely ignored, having few systematic discussions about the externalities imposed on society derived from potentially harmful organizational practices (Pfeffer, 2010). According to Ferrão (2014: 10), "[the] financial crisis of 2007/8 and its subsequent spread to the economic, social and political spheres came to give a new impetus to the need to rethink the future. Hitherto relatively marginal ideas produced from fragmentary form begin to be integrated into discourses that seek to assert themselves as alternatives to prevailing perspectives in recent decades." It is in this sense that, according to the author (2014: 9), arises the need to "suggest alternative paths, desired futures, new guiding principles for economies and societies of the future," becoming aware of Fourastié's maxim: "The future is not foreseen, it is built." From this stems the interest of the present chapter, starting from the premise that organizations impose social costs through the way they manage their employees, and must do their utmost to ensure that costs are transformed into socially added value, that is, improving the quality of life of individuals and the environment where their activities take place.

We seek to weave a theoretical reflection around the influence of social, economic, financial, environmental, and other external forces in HRM approaches. We intend to launch the discussion in which HRM policies and practices can be based on the vector of sustainability in order to contribute to the improvement of social and environmental performance of organizations, adding value of social relevance.

Therefore, we intend to focus on possible specific HRM approaches, policies and practices that support the shift in the organizational management paradigm around social issues. We developed the reflection by challenging existing theoretical frameworks and policies and practices, for example, by parallelism with strategic human resources management (SHRM), bringing to the debate alternatives that can provide useful inputs to create a sustainable organizational culture. In a methodological perspective, we start by filtering and reviewing the most relevant contributions in the subject sometimes resorting to interdisciplinarity with other areas of knowledge. This chapter is assumed, therefore, as a theoretical contribution.

In a first moment, a general picture of the current context where the organizations' activity is developed will be drawn. Subsequently, and based on "Social Pollution" concept (Pfeffer, 2010), the social impact of harmful HRM policies and practices will be addressed. To understand this, we began to emphasize the economic valuation mentality that prevails in the contemporary context, placing profits and other indicators of economic efficiency above human and social considerations. A situation persists, as there are relatively few social sanctions to reverse and remedy such practices. This, despite the increasing awareness and activism of civil society, increasingly committed to bringing to the public attention its true economic and social costs.

With this framework, it will be defended the possibility to develop organizations economically well-succeed who can also be guided by the sustainability vector in terms of social effects. Therefore, it will be developed, in a third section the possibility of linking concepts of economic and social efficiency. In a fourth section, we will look at a (re)conceptualization of the source of competitive advantage for organizations, proposing a symbiosis between resource and institutional capital.

The exhibition will continue with what we envision as relevant traits in context organizations (and how these situations may also mark their future configuration). We will focus on arguments that we consider most prominent in the design and implementation of HRM policies and practices that will be transmitted to them. We will complete the reflection ironically with the identification of another question, inviting to the reflection of to what HRM models are we moving toward—sustainable human resource management as a possible replacement for the SHRM paradigm (?). In this sense, we will then present a proposal for a theoretical model that guides this reflection (Fig. 1).

The analysis will be finalized bringing to the discussion the concept of circular economy, debating if it will be a (re) orientation in the political and economic discourse. In this regard, we suggest, based on the contributions of Dryzek (2005) and following the work of Ferrão (2014) that nowadays coexist two main discourses of change: the green growth economy and the welfare economy.

2 A Present Scenario, a Legacy for the Future

Last decades have witnessed the intensification and deepening of substantive changes in the dynamics of international capitalism. These are evident in the globalization of markets, their increasing integration, in the relocation of production, in the multiplication of products and services, the affirmation of the paradigm of technological automation and robotics, the tendency toward conglomeration of companies, the change in the typology of competition, in triggering inter-industrial cooperation based on strategic alliances between companies and broad subcontracting networks. In addition to these factors, it is also important to rethink the strategies to increase organizational competitiveness, highlighting the intensification of the use of information technology and new forms of leadership and work management (Coutinho, 1992). These, according to Deluiz (2017: 73), “are some of the signaling elements of structural transformations that shape economic globalization,” which indelibly marks the contemporary context and impact on the future.

This process also invades the political, social, and cultural dimensions. It triggers changes in the restructuring of the labor market and new forms of work organization. There is growing precarious jobs, cyclical and structural unemployment, driven by the exclusion of workers from the formal market—derived mainly from *technological unemployment* phenomenon (Barley & Kunda, 2006; Oliveira & Holland, 2017). For these reasons, Deluiz (2017: 73) systematizes that "[to] economic globalization corresponds, therefore, the globalization of the world of work and the social question."

The argument of Pfeffer and DeVoe (2012) and Vohs, Mead, and Goode (2008) allows us to verify that the current prominence of language and economic assumptions, as well as the belief in markets as logic of decision, emphasizes individual needs over social needs, competition over cooperation, leading to neglect social consequences in relation to the economic ones. Indeed, and according to Bento (2011: 97), "as a consequence, a collective way of life has been established in many societies based on a level of well-being superior to the contemporary resources available and temporarily sustained, through financial leverage, by the future welfare mortgage." This "imposes a forced readjustment of the intergenerational contract" (Bento, 2011: 97), given the excessive use of resources by current generations, endangering their readiness to meet the needs of future generations. Starting within this framework, the models of economic growth and development were shaken, emerging the need to rethink and foreseeing them with different values.

It is in this sense that a purely economic/financial framework leads to the assessment of social arrangements (only) in the same terms, parsimonious and limited perspective, possibly counterproductive (Ferraro, Pfeffer, & Sutton, 2005).

2.1 Impacts of Organizational Management on People—"Social Pollution" (Pfeffer, 2010)

The dynamics of economic and financial globalization solidified the dominance of capitalism as a model of economic and financial organization (and, given its consequences, also political and social). The configurations that we live tend to translate the idea that efficiency will be achieved through the abolition of markets restrictions, which means that government intervention or market regulation, including labor markets, will be counterproductive and harmful. The corollary of this situation conveys, as a thesis, that profit-boosting organizations will deserve privileged treatment, at the same time that policies and practices developed by them will be conceived as correct, operationalizing efficiency, everything been allowed to them in order to maximize performance (Pfeffer, 1998). A clear example is that found in HRM models.

The dependence on unrestricted labor markets and the consequent erosion of the workers' connections with their organizations, together with the reduction of protections provided by trade unions and/or the government, have become widespread (Davis, 2009) as countries seek to import the market unregulated approach from the (financial) market to the labor markets. Thus, it is possible to verify HRM practices, such as layoffs and restructuring, absence of medical benefits, sick leave or even paid leave, long working hours, bullying, and other type of abuse (Rayner, 1998). Such practices bring serious harmful effects on the physical and psychological well-being of workers (Price, 2006), but also on the collectivities where they are inserted, right away their families (Pfeffer, 2010). Despite the current widespread recognition of this situation, the adoption of measures to make impossible their harmful effects still fall short of expectations. This is because this imitation and adoption of regimes that end up not protecting workers happen although empirical evidence does not solidly support an association between the "flexibility" of the labor markets thus generated and the putative benefits in terms of organizational performance, such as boosting employment, economic growth, and increasing productivity (Howell et al. 2006). For example, research on organizational downsizing shows that it does not increase the productivity of the organization as it often disrupts networks of pre-existing relationships, which makes harder innovation and other activities such as placing products on the market that depend on collaboration in units. Productivity cannot either be fostered in this way, as downsizing favors the increase of fear and distrust in the workplace. Empirical evidence also shows that downsizing has no positive effects on profitability of the organization because it does not increase the stock price, particularly over a period of two years after the event (Guthrie & Datta, 2008).

The invisibility of harmful HRM policies and practices' consequences means that the adverse social consequences arising from these can unfold. Although organizations know what to do to make their workplaces more efficient and sustainable, relatively few actually undertake concrete actions, which highlights the urgency and complexity of calling to the discussion social aspects of organizational management practices. Therefore, it is important to understand why inefficient management practices persist, as the end of "Social Pollution" (Pfeffer, 2010) created by toxic work practices makes a socioeconomic sense not only for society but also for organizations themselves. According to Pfeffer (2010), a possible reason for the persistence of such paradoxical practices—and for externalities—may lie in low social sanctions (and ineffectiveness). In addition, it is important for organizations to realize that no unrealistic commitments are required between *doing well* financially and *doing good* for the workforce and society.

2.2 *Economic Efficiency and Social Efficiency—Utopia (?)*

A pattern of development that equally and simultaneously favors economic, social, and environmental aspects, creating synergies, is increasingly necessary and urgent. Notwithstanding the interest of the theme, it appears, right away, its complexity, given that HRM (re)orientation toward sustainability brings changes in the structure, competitive priorities and set of values chosen by organizations. For an increase in social welfare, technological progress and economic efficiency achieved by productivity increases will have to be reflected in an income redistribution, that transfigure into social investment by governments to compensate for losses during the process (such as those derived from technological unemployment, considered "collateral damage" in some speeches).

According to Wilkinson, Hill, and Gollan (2001), two issues are crucial in this sustainability debate.

The first addresses the changes that contemporary (and future) organizations need to undertake to become agents of sustainable development, providing an effective contribution to the implementation of a development model that meets the needs of today's generations without harm the future ones. The second corresponds to the search for a more ethical and holistic management model (right away of HRM), which stimulate staff development, help achieve the organization's sustainable goals and leave a positive footprint in the development of the subsystems to which it belongs (Hart & Milstein, 2003).

It is convenient to deal with these issues in an integrated and synergistic manner, not in a logic of trade-off, as they require a change in the organization's performance appraisal standard (Svendsen, 1998). It is in this sense that the relevance of the *economic efficiency–social efficiency* debate emerges, which we want to develop by trying to explore the possibility of synergy. This proposition is based on the argument of Oliveira and Holland (2017: 75), suggesting that the interconnection of the two concepts will require an inversion of the productivity principle: Social productivity must be based on the opposite of economic productivity, attainable through transformational leadership. The authors make a very relevant caveat, which reflects the complexity of the discussion: warn that the words *efficiency* and *economy* are often juxtaposed, but not so sharp between the words *efficiency* and *society*. Effectively, an efficient market is based on economic criteria, competitive advantage, and private gain. Therefore, the operations that take place in will be aimed at satisfying consumer needs and preferences, so the focus will be on market innovation. From its side, an efficient society is concerned with and based on social criteria, mutual advantage, and social profits. The operations will then be addressed to the satisfaction of social preferences—education, health, environmental preservation, and promotion of the quality of life. Therefore, the focus will be on social innovation, in other words, innovations of processes and products/services observed in the above sectors.

This argument is pertinent because it implies a logic of complex interdependence, which, however, is currently confused and distorted. This is because at present it tends to base the notion of *social efficiency* into *economic efficiency*, the latter being grounded in a pressing concern with the satisfaction of *markets* over *people's* satisfaction. However, it turns out that *people* are essentially the agents of construction, development, and vitality of the economy.

It has been assumed that social welfare could be achieved only if an economy became more efficient—*competitiveness* should precede *welfare*, *economic growth* would be a precondition for *income redistribution*. However, this logic has flaws because it tends to not recognize (or at least not properly) the intrinsic circularity of expenditure and income. In this sense, we suggest that economic efficiency should *proceed* rather than *precede* social welfare, given the argument that income redistribution supports (not drains) economies.

The interconnection of these two concepts may be weaved by the notion of *social capital*, which captures the idea that bonds and social norms are fundamental to sustainability, assuming an ideal of *social cohesion*, being implicit the concept of *cooperation*. In this regard, four characteristics will be important: relationships reliable; reciprocity and exchange; rules; standards and common sanctions; and connection in networks and groups. Trusts lubricate cooperation and thus reduce transaction costs between people. This requires common rules, standards, and sanctions to ensure that all stakeholders take enlightened and equal share in the activities, which will eventually foster reciprocity and exchange between them, with positive effects on the connection of networks and groups (Coleman, 1988; Wade, 1994).

2.3 Sustainable Competitive Advantage—Symbiosis Between Resource Capital and Institutional Capital

Business world is becoming increasingly global and demanding, so organizations are driven to look for (new) ways to withstand fierce competition and improve their performance. The most critical challenges they face are the need to increase productivity, improve organizational capabilities, expand into global markets, develop and implement new technologies, respond to demanding and volatile customer needs, adapt to a global and highly unstable market, increase revenue and decrease costs, attract and retain high performing and flexible workforce, introduce and manage relevant organizational changes, among others (Burke, 2005).

To act as a potential source of sustained competitive advantage, Barney (1991) argues that a resource must have some attributes: be valuable, rare, inimitable, and not replaceable. It should also be relevant (Grant, 1998) and dynamic (Johnson, Scholes, & Whittington, 2008).

The resource-based approach highlighted the critical role of HR in establishing and sustaining competitive advantage. However, there is an ongoing debate about what in particular provides value to the organization—the HR per se (Wright, McMahan, & McWilliams, 1994), or its management (Pfeffer, 1994; Becker & Gerhart, 1996). These lines of analysis may be interconnected, as the scope of competitive advantage in human capital organizations will be taxable on the possession of value-added human capital, requiring a set of HRM processes and practices, that is, it is necessary the existence of unique, inimitable, and transferable HR, and its change into a core competency/organizational capacity through the establishment of an HRM system (Grant, 1998; Kamoche, 1999).

Nevertheless, we suggest the need for greater complexity and rigor when looking at this debate in the current economic, political, and social context. This, since it is noteworthy that the value of HR depends not only on the organization or industry concerned, but also on certain national factors, as political, economic, educational systems, etc. Thus, and not denying the value of the theoretical approaches hitherto developed, we suggest the relevance of adopting a more holistic perspective as far as this point is concerned. In line with Ma (1999), we consider that a single source of competitive advantage may not always be sufficient to achieve superior organizational performance—various sources must be interrelated. In line with Mueller (1996), we then argue that management practices per se cannot turn HR into a valuable strategic asset because competitive advantage (and its sustainability) can only be achieved by combining HR with its management, this last irretrievably integrated into a holistic conception of the environment in which organizations operate.

This competitive advantage will therefore be the result of "social architecture"—from social standards constants in a given development process, difficult for competitors to replicate because of their evolutionary nature, long and undefined. According to Barney and Wright (1998), the organization will thus be able to create and capitalize specific (rather than general) assets achieved through HRM systems (not just by loose practices), through the work of teams (rather than individuals), recognizing economic and social HRM consequences, understanding (and capitalizing) the role of the HR function in the creation of future organizational capacity.

Only in this way will the organization be able to capitalize on isolating mechanisms, issues of resources that prevent that other organizations can get and replicate (Mahoney & Pandian, 1992). Examples include tacit knowledge and abilities, unique, invisible, complex, or dependent on the pursuit of a certain path (Lippman & Rumelt, 1982). These mechanisms ultimately explain resources mobility barriers, arising from the inability of organizations to acquire and imitate resources (Oliver, 1997).

While resource-based theorists assume that managers make rational choices constrained by uncertainty, information asymmetry, and heuristic biases, institutional theorists assume that managers often make non-rational choices bounded by social judgment, historical limitations, and inertial force of habit. In this way, the organization will be able to balance economic and normative rationality, as this conceptualization of the source of competitive advantage can connect the resource-based approaches on those with an institutionalist nature. Building the organization source of competitive advantage in the inimitable and non-transferable abilities of their HR—well managed through processes and practices of a conscious management team, inserted in the specificity of the economic, political, and social context where the activities take place—the organization will be able to capitalize on the environments specificities, where develops activity, and achieve additional income, as it will be possible to balance economic rationality (motivated by efficiency and profitability), and normative rationality (choices induced by historical precedents and social justification) (Colbert, 2004; Kazlauskaitė & Bučiūnienė, 2008).

The sustainable competitive advantage of an organization will depend, in this logic, on its ability to manage the institutional context of its resources decisions, which includes the organization's internal culture, broader influences of the State, the society and identity relations that define a socially acceptable economy. Thus, and according to Oliver (1997), we suggest that *resource capital* and *institutional capital* will be indispensable for the sustainable competitive advantage of organizations in the current and future economy. *Resources capital* encompasses the assets and value competencies of the company; *institutional capital* involves the company's ability to support assets and competencies that enhance its value proposition by with respect to the context surrounding resources and resource strategies that increase or inhibit optimal use of valued resource capital. To *resource capital*, the main success factor will be the protection and acquisition of inimitable resources. To *institutional capital*, the main success factor will be the effective management of the company's resource decision context.

Together, the factors that support and deplete *resource capital* and *institutional capital* imply ideal structural configurations in organizations, which include decentralized structures, incentive systems that reward innovations, operational levels built on multidisciplinary teams to facilitate apprenticeship, formal assessment systems, horizontal technology and information flows, as well as selection and development programs that emphasize the experience and knowledge of human resources (Oliver, 1997).

In sum, the relevance of this perspective is based on the fact that it embraces, in a logic of synergy, issues hitherto perceived as conflicting—the material dimensions (physical resources and financial performance) and immaterial (HR and social performance). This perspective evolves and corrects others, while challenging them, namely those that sustained the source of organizational competitive advantage in technologic superiority at products or processes level, in strategies and operations of marketing or advertising, or even in financial performance (capitalization).

3 Future Organizations—What HRM Policies and Practices?

The argument that has been made emphasizes *organizational sustainability* as imperative in the management models adopted. Expanding the reflection of Barney and Wright (1998), organizations should be able to answer questions such as,

- What is the organization current core competency? What is the core competency that the organization seeks to develop in the next 5–10 years?
- What is the organization current source of competitive advantage? Will it fit in with the future context?
- What will be the competitive scenario of the product/service markets and the labor markets of the organization 5–10 years from now?
- The current HR of the organization are adjusted to the maintenance and development of its core competency? What type of HR does the organization need to compete in the next 5–10 years?
- What strategic partnerships can the organization build to face current and future opportunities and/or threats?
- What kind of risks does the organization currently face? What kind of risks could it expect in the future?

- What kind of HR practices will be needed today to solidify the organization in the future?

It is then necessary to formulate HR strategies that stimulate, interconnected, the organization economic, social, and environmental performance. This perspective is reinforced by Eisenstat (1996), who states that the HR function has a central role in organizations and can stimulate the inclusion of sustainability issues in the various relationships that occur within an organization, between organizations and with external organizations; a process, in Wright and Snell (1991) optics, interactive and dynamic.

Relational management models can be crucial as they enhance the workers' involvement in the organization's decision-making process, getting their commitment and hence favoring the achievement of high performance levels through concern for the well-being, of both suppliers and users of the products/services. It is in this sense that we present some clues that could guide HRM policies and practices so that organizations can be able to face the contemporary challenges and those foreseen in the future.

(a) **Transformational leadership**

Employees tend to have implicit and/or explicit expectations about an organization. Kahn (1990) systematizes three psychological conditions that induce commitment: People tend to take part in contracts where they see clear benefits (psychological significance), protective guarantees (psychological safety), before whom they believe they have resources to honor (psychological availability). Given this win-win logic, we suggest the relevance of this type of contracts in contemporary HRM models, as they achieve a balance of interests between the different stakeholders of the organization, satisfying them in an interdependence dynamic, where a vector of mutual advantage is clear.

It is following these contributions that we can allude to the concept of transformational leadership and its relevance in current HRM models. Bass (1999) exposes the concept, pointing out three factors in how this can encourage workers to trust, admire, and respect the leader: (1) increased workers' awareness regarding the task relevance, (2) workers' persuasion to focus on the team and the organization goals, linking them together, and (3) recognition and facilitation of the workers' personal ambitions. Therefore, organizations will need to be willing to learn through tacit knowledge and capabilities implicitly acquired at operational levels, not falling into temptation to take refuge blindly in a prior determination of organizational aims, considering what should be known and done only on the basis of managers' view, regarding what they conceive as the best management method (Oliveira, 2007).

It is therefore crucial to realize that psychological contracts need to be reviewed whenever contingency changes, paying particular attention to the performance evaluation and management processes, as well as the existing conditions in the organization for meeting the established expectations (Parzefall & Hakanen, 2010).

In practice, most upper-level managers still assume that organizational change will take place per se, understood merely as a parallel value, being its meaning and dynamics transmitted to the various subsections' managers. They do not conceive the need to ask suggestions from workers or even to mid-level managers for improvements in operational practice, which could have relevance both in the value creation for the products/services they offer, and for the organization efficiency as an all. Nor do they recognize that through reflective practice one could open the way to new capabilities for individuals and work teams, new innovation' trajectories to the various operational units, with broader learning and experience implications to the entire organization (Schön, 1987).

(b) **Partnerships**

The establishment of alliances and the development (deepening and breadth) of relations with partners will be another crucial bet, by organizations, to meet the challenges of the current context, given that this raises complex obstacles for which organizations understand do not possess, only by themselves, the resources needed to meet them. On this path, it may be defensible that future HR work will develop in organizations that differ greatly from what we have today.

It is also noteworthy that achieving social goals often requires cooperative alliances between organizations and other agents along the accrual chain. In collaboration with partners, organizations may go beyond their scope of action, possibly influencing governmental actions, address shared research and development needs, expand access to new markets, among other areas, in a degree of complexity and growing ambition. In this sense, it will be possible not only establish occasional collaborations, in specific fields of action, but also undertake projects with increasing resource commitment and greater risk exposure, launching into mergers, acquisitions or joint ventures (Jackson, Renwick, Jabbour, & Muller-Camen, 2011).

Collective resource management programs that seek to build trust, develop new standards, and create training groups have, therefore, become more and more common, being designed, in Pretty (2003: 1914) words, by the terms "community-, participatory-, joint-, decentralized-, and co-management." For example, congruence of HRM systems by all alliance partners may be useful to stimulate and reinforce a consistent set of environmentally and socially friendly behaviors, as well as create a consistent set of metrics to measure them.

However, managers/employers rarely evaluate HRM systems of potential alliance partners to assess how well they fit into such management models (Jackson et al. 2011). Nevertheless, Flora and Flora (1993) report that most of these joint ventures continue to be successful, with positive financial, economic, ecological, and social results.

(c) **Think global, operationalize local**

In 1997, Moran, Harris, and Stripp predicted that (a) global organizations would become almost indistinguishable in terms of national characteristics, as they would be formed by people, from all over the world, even if their functional body of workers had the same composition. And (b) that the strategic alliances building increasing dynamics, along with the implementation of localization systems, marketing and global distribution and distribution, would cause integration and synergy between economies. This reflection allows us to point five main sources of homogeneity among organizations, acting both on their institutional and resource capital[1]: regulatory pressures, strategic alliances, human capital transfers, social and professional relationships, and competency plans (Zukin & DiMaggio, 1990). So, it is generally agreed that globalization has evolved from a key concept to a reality.

Studies indicate that HRM systems around the world are already becoming more similar, converging on data values (Brewster, Mayrhofer & Morley, 2004). These sources of homogeneity result from the incorporation of an organization into influential political, economic, and social relations—with the government, business partners, consultants, competitors, professional associates, among other sources—and are exercised by such agents, which define or prescribe socially acceptable economic behavior (Zukin & DiMaggio, 1990).

This situation is observed given the growing tendency for organizations to aggregate into cooperation schemes, being "contaminated" by the HRM policies and practices practiced by its partners. Indeed, adopted HRM systems have evolved in response to converging institutional pressures to effective social and environmental management, marked by the imperative of sustainability, which encourages not adopt a single global HRM system, but that the strategy focuses on the dependence of multiple HRM systems that reflect local idiosyncrasies. This is because, despite the globalization homogenizing forces, we are simultaneously witnessing the permanent reconstruction of heterogeneity and fragmentation via new inequalities and recreation of difference, triggering a crisis of paradigms (Haesbaert, 2009).

The place seems to (re)emerge as a revaluation of the singular, the difference, in some ways acting as a counterpoint or a strategic solution to globalization (Haesbaert, 2009). Acting locally allows, this way, that the operational context, its resources, threats, and opportunities are better known. Conversely, in thinking and acting globally, rules of a higher order than traditional scales will be set, implementing means to apply them.

This awareness is useful in the scope of HRM models as, sometimes motivated by the desired presence and active action in international/global circuits, organizations end up implementing human capital management policies and practices with a homogenizer nature that may not produce the expected results, as they do not pay attention to the specificity of each of the workers. So, this *think global*, *operationalize local* approach will, in our view, emerge as a warning and possible solution for this situation, as it will allow for the "synthesis" that gives the perception of cohesion/coherence, involving the multiple dimensions of the space where the operations take place, above all not neglecting the human dimension for the pursuit of economic and financial efficiency.

Drawing on the contributions of Moran, Harris, and Stripp (1997), current and future organizations should act in the following three principles:

[1]The concept "capital" as used throughout this chapter refers to a durable resource or capacity, but not necessarily tangible, which produces services throughout its useful life, which contribute as a source of sustainable competitive advantage for a given organization (Oliver, 1997).

1. understand macroeconomic issues, with particular emphasis on social and demographic ones, and react by adapting operational strategies;
2. fill in cultural gaps, in order to communicate more clearly and effectively, increasing the capacity for local and abroad success; and
3. adopt a global spirit, that enables it to operate effectively in the new scenario of the economy and world politics.

(d) **Equating the interests of all stakeholders—the relevance of benchmarking**

Conventional HRM models focused on establishing the economic value of HR systems, elevating investors as stakeholders of primary interest. However, it is unlikely that if employer–employee relations are merely characterized by market linkages, organizations are willing to share confidential information, invest in people and/or delegate authority to them, training them and developing skills to combine organizational with individual development (Pfeffer, 2010).

In the present context, we suggest, according to Ulrich (1997),[2] that balancing the needs of the diverse public raises questions about HRM focus. This is because each public presents merit, reason why HRM policies and practices can (and should) simultaneously focus on improving the workforce (ensuring competent, committed and dedicated employees), serving external customers (creating organizational abilities that they value/pay), investors (taking cost-cutting actions, fostering profitability) and the government (developing policies and implementing practices of national interest).

We suggest that the organization's ability to promote the satisfaction of various stakeholder interests can be achieved by benchmarking, by exploiting the concept of "*co-creation spaces*" (Coutinho & Pereira, 2010). It should be noted that this is only achieved if mental attitudes are fostered encompassing *motivation, skills,* and *abilities* to identify *opportunities* and realize the production of a new value. Coutinho (2003) clarifies such concepts by explaining their dynamics: *Motivation* is defined by the ability to catalyze new cultural, economic, and social performances; *capabilities* are seen by the competencies, aptitudes, and skills that build the potential for this performance; *opportunities* are gauged by the resources, projects, and commitments that catalyze human capabilities for the development. This way, it will be possible to elaborate a more rigorous, ambitious, and complex analysis, given that the organization looks at the challenges it faces, including in the analysis what is beyond economic and financial sustainability, managing threats, encompassing also the social domain.

Therefore, the design and implementation of well-informed HRM models will enable organizations as "spaces for co-catalyzing opportunities, where each individual is a co-creator of solutions and not a mere executor of tasks and functions"

---

[2]Already in 1997, anticipating the paradigm shifts that would come during globalization, Ulrich warned to the need for organizations rethink their strategy and actions toward the target audience. He organized the debate by questioning the dialectic of the workforce versus client versus investor versus government, which had hitherto guided organizational strategies and practices, proposing that the four dimensions were (and should be) reconcilable.

(Coutinho & Pereira, 2010: 12). Sustainability will be exploited simultaneously as "a support of the economic, financial, social and political dimensions," given the conceptualization of competitiveness "as a dynamic space where they materialize and provoke economic processes, integrating an interactive space that constitutes the feeding artery of sustainability" (Coutinho & Pereira, 2010: 12).

In order to achieve such dynamics, one of the vectors that deserve attention will be the promotion of knowledge transfer among stakeholders. This means that best practices will be shared within organization, between organizations and between them and external agents (government, business partners, trade unions, environmental associations, etc.). This will ensure the involvement of all stakeholders in the decision-making process, also fostering the achievement of better levels of performance in the various domains affected by the organization's operations, producing a more sustainable footprint. It is in this sense that benchmarking actions assume relevance as, through conferences, publications, consultants, and other forums can be quickly discussed and shared ideas, leading to new reality perspectives, alternative solutions to problems and even the prospecting of potential threats and opportunities in a given operational context (Ulrich, 1997).

This perspective is relevant since it views reality differently, because attractiveness, productivity, and worker performance ultimately depend (simultaneously as a departure and arrival point) of one's own development and social cohesion (O'Gorman & Kautonen, 2004). The interaction and cooperation between different institutions should be placed at the service of the organizational sustainability, in an integrated approach, "with planning strategies aimed at tune the economic aspect with the social aspect" (Coutinho & Pereira, 2010: 4).

(e) **Evaluate and capitalize on the results of this evaluation**

According to Ulrich (1997), HRM practices, to be efficient, will need to evaluate more, more accurately, with scientific rigor and demanding. This, in order to effectively be possible to relate HR's contribution to business, financial, environmental, and social results of the organization. So, this will require not only the development of objective, valid, measurable, and reliable measures of HR effectiveness, as well as broader assessment models, looking at impacts of the HR contribution in all areas where the organizational activity develops repercussions. It will be also necessary to enter these results in a systematic co-creation logic in the future review and planning of organizational strategies and practices, capitalizing on the good points and reversing the negative aspects. This dynamic may also have underlying the publicizing of such results.

Fletcher (2001) leaves some warnings as to what may be a non-objective evaluation process and therefore inefficient in its results:

- the belief that the evaluation process could have negative impacts on the motivation and corresponding performance of the worker,
- the desire or need to protect workers whose performance has been reduced due to personal problems,

- the absence of confrontation or conflict with the potentially most problematic subordinates,
- the minimization of payments for the achievement of certain performance levels,
- as well as the encouragement to exit of unwanted subordinates.

A possibility of reconverting these negative aspects may lie in a dual and simultaneous process—subordinates evaluating their peers and superiors, and superiors evaluating their peers and subordinates. This process needs to be undertaken on a dyadic rather than bilateral basis because it can be criticizing the fact that it contains aspects of potential doubt or conflict, such as some subjectivity in the assessment peer review, confidentiality issues, and results advertising. This model could provide useful contributions in the achievement of higher management levels, indicating the practices that may be dysfunctional in the organization, rather than merely referring to whether or not workers have complied with certain performance levels, which can sometimes be the cause of the highlighted dysfunctions.

In this sense, effective mutual feedback is vital for both workers and employers have an active voice not only in the defense of their own interests but also in the pursuit of organizational objectives and their revision when necessary. These voices should articulate each other to objectify which is, otherwise implicitly assumed in the organization (Oliveira & Holland, 2017).

We suggest that, this way, it will be possible to overcome the problem based on the fact that the operational logic is top-down designed, without concern for learning at the base operational level. This is related to the need to identify and evaluate proximal and distal processes and outcomes (Peccei, Voorde, & Veldhoven, 2014). The former relate more closely and directly with HRM practices (involving absenteeism, turnover rates, productivity, quality and performance of the service). The seconds relate the overall financial or market performance of the organization and may, at a first appear, seem to be less correlated directly with HRM practices (involving return on capital or in the market value, in case of a private sector organization, or the total expenditure on a public service). However, an interdependence between the two vectors appears.

(f) **HR centrality in innovation—the specialization of functions**

For decades, HRM has focused primarily on achieving results/performance through the programs design that affected people and work processes. However, more and more emphasis should be placed on processes and the outputs themselves, not only at the organizational level (commercial, financial, etc.), but also workers (skills developed, new knowledge acquired, lived experiences, etc.). These two vectors are interdependent, creating a circular dynamic.

We suggest that multiskilling and multitasking may not be effective practices. This is because, according to Adler and Benbunan-Fich (2012: 159), 'Multitasking situations present a resource-allocation problem to humans because they must decide how to assign limited resources (such as time or attention) across multiple competing tasks to achieve specific performance goals." This will be because the simultaneous work in different tasks favors the retention of temporary information, which is only useful for the development of that task at that particular moment (Borst, Taatgen &

Van Rijn, 2010). In addition, the constant volatility between competing tasks ultimately distracts the worker and thus degrades his performance (Rubinstein, Meyer, & Evans, 2001), as it results in slower response times as well as in higher error rates (Payne, Duggan, & Neth, 2007). Such costs, according to Yeung (2010), arise from the interference between-task, which results from the residual tendency for the worker to keep thinking on the task recently abandoned and currently irrelevant.

Alternatively, we suggest that skill profiling and skill path planning may foster personnel development. Given that it is this asset, the human capital, that better knows the tacit rules and implicit norms that may be potentiating or make impossible the organizational efficiency and learning (Gupta & Singhal, 1993). In addition, horizontal mobility may give workers the opportunity of tasks and job variation, alleviating the intensive front-line labor pressure (Oliveira & Holland, 2017). It is in the course of this argument that we allude to the contribution of Schneider and Bowen (1993), who argue that "[t]he motto is to manage human resources contingently."

(g) **Variability of incentives**

Incentives and rewards are often considered to be the most powerful mechanisms in the connection of organizational interests to those of employees. We therefore suggest that paying attention to compensation schemes that link incentives and rewards to performance, and combine them with social objectives, will be a relevant bet on HRM models in the current paradigm. This is because they will be seen in a logic of mutual interest, creating a win-win-win dynamic: balancing the interests of the employer, worker, and society (Jackson et al., 2011). In this regard, strike a balance between life and work, through the possibility of customizing individual work time, could act as a crucial motivating issue, and even as an alternative to the monetary incentives, frequently applied in an exclusive way (Oliveira & Holland, 2017).

However, the development of effective incentives (early monetary) can be a challenge given the difficulty of assess accurately and fairly behaviors and respective social impact (Fernandez, Junquera, & Ordiz, 2003). In this regard, Jackson et al. (2011) give an alert, stating that organizations should strive to strike the right balance between the use of motivational mechanisms and the achievement of their results, because if the punishment for poor performance in social goals is considered too harsh, managers/employers may withdraw from involvement in sustainable management, an aggravated factor if weak rewards are perceived.

(h) **Management of and in diversity**

More and more organizations are striving to develop HRM systems that fit the contours a complex, multi-faceted, volatile, and ambiguous international landscape aimed at addressing the challenges and as far as possible capitalize on them (Pinskse, Kuss, & Hoffmann, 2010; Tarique & Schuler, 2010).

Right away, we highlight *the challenge of diversity management* and *the challenge in management diversity*. This will bring implicit the consolidation of a culture of inclusion, defined as an organizational context that allows people with experiences and different mental models cooperate to achieve individual, organizational, and collective growth (Pless & Maak, 2004; Wimbush 2006). Taking this into account,

Wilcox (2006) defends that management of and in diversity may involve defining an HR policy for employees who belong to groups traditionally marginalized.

According to Fleury (1999), organizational performance will be highly taxed from the management model of diversity adopted in the organization because it:

1. attracts and retains the best employees from among those who seek a position in the labor market;
2. provides access to market share previously untapped;
3. stimulates creativity and innovation;
4. creates potential for the resolution of complex and multidimensional problems; and
5. increases organizational flexibility.

(i) Technical and technological progress—(hostages of) the automation paradigm?

In the contemporary context, information systems (IS), based on the intensification of the use of information and communication technologies (ICT), are crucial to the success of organizations. They emerge, then, new tendencies toward work: Work becomes more abstract, intellectualized, autonomous, collective, interdependent, and complex. For an organization to be able to meet such a challenge and obtain, from it, competitive advantage, is urgent to capitalize on technical and technological progress. Adapting Deluiz's reflection (2017: 74), this will pass through the awareness that "the symbolic interposes between the object and the worker. The work object itself becomes immaterial: information, 'signs', symbolic language."

There exist, thus, different concepts of production, which bring organizational changes and thus the own redefinition of base vectors, such as the notions of *profile, competences,* and *tasks*. This, according to Deluiz (2017: 74), given the "unpredictability of situations in which the worker or the collective of workers have to make choices and options all the time, expanding the mental and cognitive operations involved in the activities, but at the same time their subjective costs." This situation highlights the need for an awareness of a radical change in the conception of employment and work. As Deluiz warns (2017: 74–75): "this is no longer a formal qualification/prescribed qualification/worker qualification to develop work-related tasks, defined by the company to the establishment of salary grades or by certification or diploma training systems, were tasks were described, coded, and could be visualized. But the real qualification of the worker, understood as a set of competences, skills, and knowledge, that come from diverse instances, such as, general education (scientific knowledge), vocational training (technical knowledge) and work and social experience (tacit qualifications)."

We suggest that the actual context gives rise to the need for clarification of the concepts of *profile*, *competences,* and *skills.* It is this way because the real qualification of workers involves "[the] set of skills put into action in a concrete work situation, the articulation of the various knowledge from various spheres (formal, informal, theoretical, practical, tacit) to solve problems and cope with unpredictable situations, the mobilization of intelligence to meet the challenges of work"(Deluiz, 2017: 74–75).

That is, such concepts, despite their interconnection, are distinct. What is, sometimes, seen in organizations is a juxtaposition or confusion between them, considering as a worker's profile only their technical qualifications and/or tasks performed.

So, the notion of *tasks* brings a markedly objective and technical nature, easily delimited by a function analysis. The notion of *competences* can link this more technical and quantifiable dimension (technical skills—academic training or other professional training type), including a more subjective dimension (social skills—emotional intelligence, teamwork, leadership, etc.), mainly the result of social experience. The notion of *profile* will synthesize these two fields, encompassing the qualification and academic and professional experience of the individual, adding to their way of being and thinking, their aspirations, desires, and motivations. Therefore, this logic of thought implies conceiving the worker in its specificity, to view him as a constantly evolving complex, similar to the process of growth and development of the organization and the context in which it operates.

Thus, to persuade workers and job vacancies implies broadening the thinking matrix beyond cognitive dimension, intellectual, and technical competences (Deluiz, 1995). Workers will be able to add value to their organizations by providing ideas, programs, projects, and business initiatives, helping them mature and become more competitive. Organizations will then be able to count with an inimitable asset. According to Ulrich, Brockbank, Yeung, and Lake (1995), the value-added reasoning that is generated is based on a causal logic of five steps:

1. business conditions are changing dramatically;
2. to respond to such turbulent conditions, organizations must allocate resources to sustained competitive advantage;
3. competitive advantage derives from the creation and maintenance of sources of uniqueness that are not easily replicable by competitors;
4. organizational capacity consists of a unique set of organizational attributes that deliver value to customers and that cannot easily be replicated;
5. HR practices are critical in building and maintaining organizational capacity.

All these points converge to a reflection: As the pace of change increases, HRM needs to face that paradigm, through internal adaptation, proportional to what occurred in the external environment, otherwise, the organization will not be able to be competitive (Ulrich et al., 1995). Indeed, it redefines the relationship between *know-to do* and *know-to be*, hitherto conceived as almost mutually exclusive, based on the organizations' competitive advantage in the first dimension (the most tangible and immediate) due to the strong financial orientation. The current trend, however, suggests that the real qualification of workers is much more complex to be materialized, as it focuses essentially on *know-to be,* thus not belittling the *know-to do*.

4 Toward Sustainable HR Management (?)

Strategic human resource management has emerged as dominant model in the last three decades. However, in the last decade we have witnessed the solidification of a new approach: sustainable human resources management. This new conception is based on the urgency of articulating an agenda for the global change and the development of a common future for Humanity, worrying about the interconnection of social and economic progress without threatening current and future living conditions. Organizations will need to create value in three areas, seen as the three pillars of sustainability—economic, social, and environmental—going beyond intra-organizational boundaries (Elkington, 2006). As a corollary, *organizational sustainability* can be understood as the managerial practice that aims the organizational vitality timeless, performing ethically plausible actions that favor equitable economic, environmental, and social performance criteria (Vickers, 2005; Jabbour & Santos, 2008). So, this concept articulates an approach in which organizations emerge as prospectors rather than reactors to the volatility of the current international context.

Jackson and Seo (2010) point out that work in the areas of HRM and sustainability shares some common elements that offer opportunities for integrating these two research streams, such as:

1. acceptance of the economic imperatives that shape managers' behavior;
2. conceptual structures shared to describe the strategies on which organizations are based to maximize its performance;
3. increasing recognition that it is crucial to address the concerns of multiple stakeholders; and
4. taking an active part in an increasingly global economy.

Current HRM and sustainable organizations require a holistic and long-term planning and action focus beyond the conventional exclusive pursuit of financial performance (Wilkinson et al. 2001). This is taking into account the awareness of the scarcity of HR itself, given the progressive aging of the workforce and related health problems. Hence, it appears the argument that the promotion of organizational sustainability (specifically in the HR subsystem) will become a strategy for survival for organizations (Ehnert, 2009). Therefore, according to Boudreau and Ramstad (2005), there are two main challenges for contemporary HR management: One is related to attraction, maintaining and developing talent needed for the survival of an organization in the globalization age; another covers the design and implementation of HRM models that meet the objectives—interrelated—of economic, social, and environmental sustainability. The authors, then, maintain that stimulating organizational sustainability is the current HRM paradigm.

Based on the above, and according to Kramar (2014: 1084), sustainable human resources management can be defined as "the pattern of planned or emerging HR strategies and practices intended to enable the achievement of financial, social and ecological goals while simultaneously reproducing the HR base over a long term." In this sense, it aims "to minimize the negative impacts on the natural environment and on people and communities and acknowledges the critical enabling role of CEOs, middle and line managers, HRM professionals and employees in providing messages which are distinctive, consistent and reflect consensus among decision-makers." So, for Ehnert and Harry (2012: 222), "[t]he sustainability debate at the corporate and HRM level deals with practices and strategies that produce significant impact on an organization's natural and social resources and environments which then influences the organization's and HRM's future management conditions and business environment."

Sustainable human resources management is, therefore, a reforming approach, challenging the conventional premise that HRM's purpose would be to achieve business outcomes. Instead, by incorporating the notion of *negative externalities* (Mariappanadar, 2003 and 2012), has as its core the premise that the development of human capital will act as the essential result of HRM processes, taking an explicit moral stance on desired outcomes in organizational practices. This conceptualization stems, according to Kramar (2014), from the fact that this approach recognizes and values, as a cornerstone, the influence of the organization institutional context in the development of HRM strategy and policies. So, organizational results (emphasizing the relevance of human and social results) will be broader and more emphasized compared to the financial ones. It is clear that its main concern involves the organization long-term survival, looking to correct some criticism of "myopia" that could be addressed to the HRM conventional vision, due to the overvaluation of the business in the short term.

Therefore, it is a model subject to a higher level of complexity, compared to that observed in strategic human resources management. At first glance, sustainable human resources management may be conceived as an extension of strategic human resources management. However, the literature documents clearly the specifics of each approach. Strategic human resources management emerged and developed in a dynamic and turbulent context, in which organizations were threatened by the changing nature of labor, given the vulgarization of part-time system and services. Globalization has fueled both competition and interdependence between organizations and fields of action, fostering concern for value creation. The core of this approach was based on the interrelation between HRM and financial performance of the organization, as it was argued that HRM practices will bring profit to businesses through people (Ulrich 1997).

The same is not true in sustainable human resources management, as this will have to be conceived as an open system.[3] The achieved results can be measure by the assessment at the organizational, social, individual, and ecological level, as this approach considers them articulated.[4] Central to all these issues is the clear focus on the purpose of HRM practices, giving it an indelible moral character, which was

[3]To more information please see Ehnert (2009) model and review developed by Kramar (2014).

[4]In Kramar (2014: 1081–1082) words "*Measures would need to evaluate outcomes such as the quality of the employment relationship, the health and wellbeing of the workforce, productivity (organisational); the quality of relationships at work, organisation being an employer of choice and being recognised among a range of potential sources of labour (social); and job satisfaction, employee motivation and work–life balance (individual); use of resources, such as energy, paper,*

not the case at strategic human resource management. It is in this sense that HRM socially appropriate requires:

1. create reward systems based on the concepts of equity, distributive fairness, autonomy, and respect;
2. ensure safety in the workplace in order to prevent accidents and respect the health of employees;
3. treat employees in accordance with the principles of respect, transparency, honesty, and long-term nature of the changes; and
4. respect the privacy of the employee and, consequently, to have respect, freedom and autonomy as organizational values (Greenwood, 2002).

This argument converges to one point: HR function can play a central role in building a more ethical organizational base that integrates social aspects into its more traditional practices.

By grounding in mutual advantage, the usefulness of this perspective is based on the fact that it has relevant implications to contain social inefficiency as a result of excessive pressure on economic efficiency (Holland, 2000). As principles, organizational culture will be guided by the reciprocal recognition and understanding, the respect for plurality, truth, integrity, and a cross-cultural point of view. It will therefore value inclusive skills such as demonstrating respect and recognition for diversity; appreciation of different perspectives and cooperative leadership style; encouraging open and honest communication; integrity and the implementation of participatory decision-making and problem-solving processes. In terms of practices, such organizations will be concerned with succession plans, specialized training programs, opportunities for teamwork, and the organization of moments that foster communication and reflection (Weaver & Treviño, 2001). However, it should be noted that this approach does not yet have a cohesive body of literature, although it is in progressive development, especially since the last decade (Bansal, 2005; Pfeffer, 2010). Even if it refers to an implicit pragmatic consensus, it is still arguable that most organizations are able (or predisposed) to create this efficiency (Ehnert & Harry, 2012).

4.1 Proposed Analysis Model

Achieving alignment between organizational strategies and practices and environmental/social management requires a deep understanding of the alternative paths that sustainable organizations can take (Hoffman, 2007). Considering that their construction requires a multidimensional vision that encourages a superior performance—but subject to a moral scrutiny—we present in Fig. 1 a proposal for a model that encourages organizational sustainability based on HRM.

---

water use, production of green products and services and costs associated with work travel (ecological). The appropriate measures would need to be developed for an individual organisation and then cascaded down to all employees using HRM practices, such as role design, performance indicators and rewards."

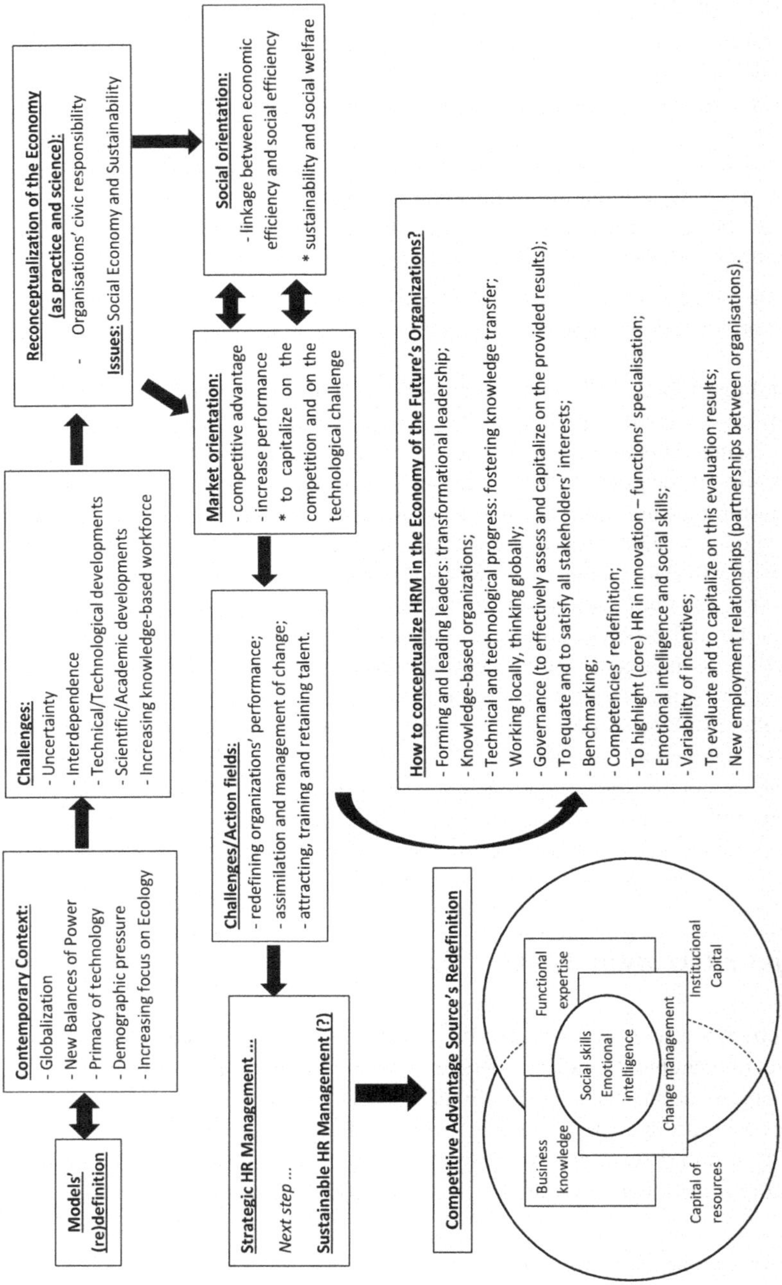

Source: Adapted and expanded from Ulrich (1997), Vickers (2005), Jabbour and Santos (2008) and Pfeffer (2010).

Fig. 1 How do we look at Organizations and Human Resource Management in the "Economy of the Future"?

The literature that advocates sustainable human resources management models represents, from our perspective, an alternative approach to current people management policies and practices, as it identifies broader purposes for HRM, by recognition of the complexities between the various actors and subsystems affected by the organizational activities. This is an inclusive and evolutionary approach, as the argument woven in no way means that organizational outcomes such as return on investment, market share, and profit are no longer relevant. We can say that its relative relevance will be reviewed given the triumph of broader concerns such as the ecological footprint and social redistribution of incomes.

In this model, we start from a critical analysis of the context in which the organization operates. It is expected that all HR practices are balanced to support organizational sustainability because the objectives of profitability, innovation management, cultural, and social/environmental diversity, cannot be contradictory. They must even be interconnected to be unified in perspective. This, since we believe that sustainable human resources management presents new opportunities for finding profitable management approaches, that simultaneously offer benefits to shareholders, employees, customers and communities, as well as the others interested. Considering that sustainable activities are increasingly recognized by employees, customers and shareholders, organizations will thereby be able to achieve a reputation which, in turn, could foster better financial and economic performance by starting a new cycle focused in HRM. This can reinforce the legitimacy of the organization itself in light of the public opinion scrutiny, strengthening its institutional image as a welfare agent that enhances the development of the environment (Jabbour & Santos, 2008).

Therefore, this model aims to provide clues for the organization not only anticipate the contingencies to which needs to be answered at the moment, but also the challenges ahead, bringing to the discussion the interconnectedness of the internal and external results obtained and the analysis of the breadth and legitimacy of interests served. Thus, organizations create an internal context that focuses on sustainability.

For that, the change management competency will be decisive, allowing HR professionals to add value to their organization by facilitating the development of organizational change capacity to encourage plasticity and the absorption of relevant volatilities in the business environment (Ulrich et al., 1995).

4.2 Circular Economy—(Re)orientation in Political and Economic Discourse (?)

This reflection exercise culminates in one point: looking at management and economy not only from a technical perspective, but imbued with a moral stamp, in a two-way and dynamic relationship. This is because the development of a society does not depend only from the potential production of its economy, the conventional interpretation of the aim of *maximizing output*: Economic and management actions are morally oriented by the values framework of the people who execute it, and this is being influenced by the context, by those actions and their consequences (Bento, 2011).

Therefore, a dynamic is created between self and collective interest. Because despite such actions being based on utilitarian values, "this system should always be understood as having a subsidiary role" in the face of higher moral imperatives (Bento, 2011: 21), which encompass the interconnection between *efficiency* and *well-being*, in a logic of *aggregate utility*, that is, oriented as the ultimate goal for *social cohesion*. In this regard, considerations arise in the context of *Welfare Economy* or *Social Economy*, complexifying the concept and scope of action of Economics and Management, beyond operational considerations and technical models. Effectively, "when the intervention produces a result where some improve without others worsen, there is no doubt that this translates into an improvement in social welfare. But even therefore, it cannot be disputed who constitutes the group of beneficiaries, especially if the same resources could have been used to improve, albeit to a lesser extent, the situation of a different group" (Bento, 2011: 58). However, the author points out: "the recognition of the need for weighting and the choice weights are the product of moral judgments and therefore depend of the value scale of who formulates them" (2011: 59), from which the conflict may derive between (economic) *efficiency* versus *equity* (social cohesion). It should be emphasized that the emphasis should not be placed solely on one of the vectors mentioned, exclusively. The attention should be given in a strategy of constant interconnection between them, pondering its relative relevance in relation to a strategic reading of the context at the moment.

It is on this path that it is justified *sustainability* as an *imperative* in equating perspectives in relation to HRM policies and practices in the future. As in an economic community, all members interact, which makes that a decision of one impact on the objectives and actions of other(s), in other words, on the welfare of these. "For this reason, the multiplication of individual decisions creates a qualitatively different result than the mere sum of the individual results," because "the multiplication of decisions individually rational does not guarantee that the aggregate result is rational" (Bento, 2011: 51). This arises given the impossibility of removing *market failures* from the equation that objectify this inability of the Economy (of the Market) per se to act efficiently, bringing to the debate the relevance of the action of a regulatory agency.

5 Final Considerations

According to Ulrich (1997: 175), "[w]hile some answers concerning the future of HR seem to be known and shared; there is more that is unknown. In these cases, the questions are more compelling than answers." It was precisely this the starting point and the arrival point of the present chapter, because with the same we realize that:

1. HRM and the concept of HR itself are under scrutiny, which we highlight as a positive issue;

2. HRM and even HR concepts as we have known them need to be re-thought in order to adapt to change;
3. this redesign will bring with it important challenges, because it will require a (re)design and a (re)implementation of competences and capabilities.

In this regard we suggest, based on the contributions of Dryzek (2005) and following the work of Ferrão (2014), that two main discourses of change coexist today: the *green growth economy* and the *welfare economy*. To guide the reflection, we present a proposal of theoretical model. The first does not break radically "with the near-continuity growth model," the "business-as-usual," but distances itself from it and acquires specificity as it incorporates environmental concerns, focused on the search for an economic growth marked by a more efficient use of resources, having as objective "a greater competitiveness of the economy." In turn, the second "takes the prospect of prosperity beyond economic growth," based on a critique to the market economy, assuming, therefore, as base vector the existence of "clear limits to the exploitation of available resources," advocating "the need of a profound change from the review of the objectives pursued" (Ferrão, 2014: 17).

In the logic of Vickers (2005) and Jabbour and Santos (2008), by presenting what we conceive as present and future of HRM we inherently refer to *sustainability* as imperative, presenting a theoretical model of support in the reflection on the challenges that may be posed to Organizations and HRM. We emphasize that the model acts only as a result of the theoretical reflection developed throughout the present chapter, lacking empirical verification to support it. It is assumed, effectively, as another contribution to the path of the Fourastié maxim: "The future is not foreseen, it is built." Therefore, we advocate that this may also serve as a basis for future empirical research, contributing to the development of this subject, because a set of propositions and research methodologies can be derive from it.

The reported conceptual framework and research serve to inform reflection on how to develop the HRM models in a context indelibly marked by social concerns. The key implication of the proposed model is that the ability of an organization to generate income from resources will depend above all on its effectiveness in managing the context (internal and external). Therefore, the design and implementation of well-informed HRM models will enable organizations to gain insight as "spaces for co-catalyzing opportunities, where each individual is a co-creator of solutions and not a mere performer of tasks and functions" (Coutinho & Pereira, 2010).

In this vein, we suggest that future research on sustainable competitive advantage should focus on in the way how resources are developed, managed, and disseminated. Efforts to identify sources of *resources* and *institutional* capital among competitors may shed additional light on the management of both to foster sustainable competitive advantage, which underlines the relevance of longitudinal studies of the resources development and deployment process. We also propose a trajectory of evolution of this approach, testing until its solidity, following the own evolution of the contemporary context. For example, further empirical studies will be need to understand how organizations can develop a group of managers capable of leading sustainability and learning initiatives throughout the process.

Some sustainable HRM practices are already used in organizations, including the balance between professional and personal life, foreseeing future availability of resources and capabilities through workforce planning as well as the estimation of the organizational ecological footprint. Regardless, these measures need to be systematically adopted and be part of a broad and cohesive HRM strategy, marked by interdisciplinarity, informed by theories that enable the overcoming of the ambiguity and based on rigorous measuring instruments and feedback that articulates actions to be results. It is indeed useful to point out that formulating the necessary transition strategies is different from mobilize agents and decision makers to trigger and manage the necessary change processes, as in this last situation occurs the influence of more delicate issues (the time interval, the spatial scale or even the constraint on succession of management/administration cycles).

References

Adler, R. F., & Benbunan-Fich., R. (2012). Juggling on a high wire: Multitasking effects on performance. *International Journal of Human-Computer Studies, 70*(2), 156–168.

Bansal, P. (2005). Evolving sustainably: A longitudinal study of corporate sustainable development. *Strategic Management Journal, 26*(3), 197–218.

Barley, S. R., & Kunda, G. (2006). Contracting: A new form of professional practice. *The Academy of Management Perspectives, 20*(1), 45–66.

Barney, J. (1991). Firm resources and sustained competitive advantage. *Journal of Management, 17,* 99–120.

Barney, J. B., & Wright, P. M. (1998). On becoming a strategic partner: The role of human resources in gaining competitive advantage. *Human Resource Management, 37*(1), 31–46.

Bass, B. M. (1999). Two decades of research and development in transformational leadership. *European Journal of Work and Organizational Psychology, 8*(1), 9–32.

Becker, B., & Gerhart, B. (1996). The impact of human resource management on organizational performance: Progress and prospects. *Academy of Management Journal, 39*(4), 779–801.

Bento, V. (2011). *Economia, Moral e Política*. Lisboa: Relógio d'Água Editores.

Borst, J. P., Taatgen, N. A., & Van Rijn, H. (2010). The problem state: A cognitive bottleneck in multitasking. *Journal of Experimental Psychology. Learning, Memory, and Cognition, 36*(2), 363–382.

Boudreau, J. W., & Ramstad, P. M. (2005). Talentship, talent segmentation and sustainability: A new HR decision science paradigm for a new strategy definition. *Human Resource Management, 44*(2), 129–136.

Brewster, C., Mayrhofer, W., & Morley, M. (Eds.). (2004). *Human resource management in Europe: Evidence of convergence?*. London: Routledge.

Burke, R. J. (2005). Reinventing human resource management: Challenges and new directions. In R. J. Burke & C. L. Cooper (Eds.), London: Routledge.

Colbert, B. A. (2004). The complex resource-based view: Implications for theory and practice in strategic human resource management. *Academy of Management Review, 29*(3), 341–358.

Coleman, J. S. (1988). Social capital in the creation of human capital. *The American Journal of Sociology, 94,* 95–120.

Coutinho, L. (1992). A terceira revolução industrial e tecnológica. As grandes tendências das mudanças. *Economia e Sociedade, 1*(1): 69–87.

Coutinho, M. (2003). *Economia Social em Portugal. Emergência do Terceiro Sector na Política Social*. Lisboa: APSS/CPIHTS.

Coutinho, M., & Pereira, O. P. (2010). A oportunidade da cidade: contingências da conjuntura e da teoria. *Urban Studies, 39*(13), 2395–2411.

Davis, G. F. (2009). *Managed by the markets: How finance re-shaped America*. Oxford: Oxford University Press.

Deluiz, N. (1995). *Formação do trabalhador: produtividade e cidadania*. Rio de Janeiro: Shape Ed.

Deluiz, N. (2017). A globalização econômica e os desafios à formação profissional. *Boletim técnico do Senac, 30*(3), 73–79.

Dryzek, J. S. (2005). *The politics of the earth: Environmental discourses*. Oxford: Oxford University Press.

Ehnert, I. (2009). *Sustainable human resource management: A conceptual and exploratory analysis from a paradox perspective*. Heidelberg: Physica-Verlag.

Ehnert, I., & Harry, W. (2012). Recent developments and future prospects on sustainable human resource management: Introduction to the special issue. *Management Revue, 23*(3), 221–238.

Eisenstat, R. (1996). What corporate human resources brings to the picnic: Four models for functional management. *Organizational Dynamics, 25*(2), 7–22.

Elkington, J. (2006). Governance for sustainability. *Corporate Governance: An International Review, 14*(6), 522–529.

Fernández, E., Junquera, B., & Ordiz, M. (2003). Organizational culture and human resources in the environmental issue: A review of the literature. *International Journal of Human Resource Management, 14*(4), 634–656.

Ferrão, J. (Coord.). (2014). *Que Economia Queremos?*. Lisboa: Fundação Francisco Manuel dos Santos.

Ferraro, F., Pfeffer, J., & Sutton, R. I. (2005). Economics language and assumptions: How theories can become self-fulfilling. *Academy of Management Review, 30*(1), 8–24.

Fletcher, C. (2001). Performance appraisal and management: The developing research agenda. *Journal of Occupational and Organizational Psychology, 74*(4), 473–487.

Fleury, M. T. L. (1999). The management of culture diversity: Lessons from brazilian companies. *Industrial Management & Data Systems, 99*(3), 109–114.

Flora, C. B., & Flora, J. L. (1993). Entrepreneurial social infrastructure: A necessary ingredient. *The Annals of the American Academy of Political and Social Science, 529*(1), 48–58.

Grant, R. M. (1998). *Contemporary strategy analysis: Concepts, techniques, applications*. Malden: Blackwell.

Greenwood, M. R. (2002). Ethics and HRM: A Review and Conceptual Analysis. *Journal of Business Ethics, 36*(3), 261–278.

Gupta, A. K., & Singhal, A. (1993). Managing human resources for innovation and creativity. *Research Technology Management, 36*(3), 41–48.

Guthrie, J. P., & Datta, D. K. (2008). Dumb and dumber: The impact of downsizing on firm performance as moderated by industry conditions. *Organization Science, 19*(1), 108–123.

Haesbaert, R. (2009). Região. *Diversidade Territorial e Globalização. GEOgraphia, 1*(1), 15–39.

Hart, S. L., & Milstein, M. B. (2003). Creating sustainable value. *Academy of Management Executive, 17*(2), 56–69.

Hoffman, A. J. (2007). *Carbon strategies: How leading companies are reducing their climate change footprint*. Ann Arbor, MI: University of Michigan Press.

Holland, S. (2000). *Innovation agreements. Paper for the portuguese presidency of the European council*. Lisbon: Office of the Prime Minister.

Howell, D. R., Baker, D., Glyn, A., & Schmitt, J. (2006). *Are protective labor market institutions at the root of unemployment? A critical review of the evidence*. New York: New School for Social Research.

Jabbour, C. J. C., & Santos, F. C. A. (2008). The central role of human resource management in the search for sustainable organizations. *The International Journal of Human Resource Management, 19*(12), 2133–2154.

Jackson, S. E., Renwick, D. W., Jabbour, C. J., & Muller-Camen, M. (2011). State-of-the-art and future directions for green human resource management: Introduction to the special issue. *German Journal of Human Resource Management, 25*(2), 99–116.
Jackson, S. E., & Seo, J. (2010). The greening of strategic HRM scholarship. *Organization Management Journal, 7*(4), 278–290.
Johnson, G., Scholes, K., & Whittington, R. (2008). *Exploring corporate strategy: Text & cases*. London: Pearson Education.
Kahn, W. A. (1990). Psychological conditions of personal engagement and disengagement at work. *Academy of Management Journal, 33*(4), 692–724.
Kamoche, K. (1999). Strategic human resource management within a resource-capability view of the firm. In R. S. Schuler & S. E. Jackson (Eds.), *Strategic human resource management*. Oxford: Blackwell Publishers Ltd.
Kazlauskaitė, R., & Bučiūnienė, I. (2008). The role of human resources and their management in the establishment of sustainable competitive advantage. *Engineering Economics, 5*(60), 78–84.
Kramar, R. (2014). Beyond strategic human resource management: Is sustainable human resource management the next approach? *The International Journal of Human Resource Management, 25*(8), 1069–1089.
Lippman, S. A., & Rumelt, R. P. (1982). Uncertain imitability: An analysis of interfirm differences in efficiency under cooperation. *Bell Journal of Economics, 13*(2), 418–438.
Ma, H. (1999). Creation and pre-emption for competitive advantage. *Management Decision, 37*(3), 259–266.
Mahoney, J. T., & Pandian, J. R. (1992). The resource-based view within the conversation of strategic management. *Strategic Management Journal, 13*(5), 363–380.
Mariappanadar, S. (2003). Sustainable human resource strategy: The sustainable and unsustainable dilemmas of retrenchment. *International Journal of Social Economics, 30*(8), 906–923.
Mariappanadar, S. (2012). The harm indicators of negative externality of efficiency focused organisational practices. *International Journal of Social Economics, 39*(3), 209–220.
Moran, R., Harris, P., & Stripp, W. (1997). *Desenvolvendo organizações globais: como preparar a sua empresa para a competição mundial*. São Paulo: Futura.
Mueller, F. (1996). Human resources as strategic assets: An evolutionary resource-based theory. *Journal of Management Studies, 33*(6), 757–785.
O'Gorman, C., & Kautonen, M. (2004). Policies to promote new knowledge-intensive industrial agglomerations. *Entrepreneurship & Regional Development, 16*(6): 459–479.
Oliveira, T. C. (2007, June). *Delving down to learn up*. Plenary Presentation to the Sloan MIT-Portugal Conference on New Developments in Management, Lisbon, Portugal.
Oliveira, T. C. & Holland, S. (2017). Economic and social efficiency: The case for inverting the principle of productivity in public services. In C. Machado & J. P. Davim, (Eds.), *Productivity and organizational management* (pp. 169–199). Berlim: De Gruyter.
Oliver, R. L. (1997). *Satisfaction: A behavioral perspective on the consumer*. New York: The McGraw-Hill Companies Inc.
Parzefall, M. R., & Hakanen, J. (2010). Psychological contract and its motivational and health-enhancing properties. *Journal of Managerial Psychology, 25*(1), 4–21.
Payne, S. J., Duggan, G. B., & Neth, H. (2007). Discretionary task interleaving: Heuristics for time allocation in cognitive foraging. *Journal of Experimental Psychology: General, 136*(3), 370–388.
Peccei, R., Voorde, K., & Veldhoven, M. (2014). HRM; well-being and performance: A theoretical and empirical review. In J. Paauwe, D. Guest, & P. Wright, (Eds.), *HRM and performance: Achievements and challenges* (pp. 19–37). United Kingdom: John Wiley and Sons Ltd.
Pfeffer, J. (1994). Competitive advantage through people. *California Management Review, 36*(2), 9–28.
Pfeffer, J. (1998). *The human equation: Building profits by putting people first*. Boston: Harvard Business Press.
Pfeffer, J. (2010). Building sustainable organizations: The human factor. *Academy of Management Perspectives, 24*(1), 34–45.

Pfeffer, J., & DeVoe, S. E. (2012). The economic evaluation of time: Organizational causes and individual consequences. *Research in Organizational Behavior, 32,* 47–62.

Pinkse, J., Kuss, M. J., & Hoffmann, V. H. (2010). On the implementation of a 'global' environmental strategy: The role of absorptive capacity. *International Business Review, 19*(2), 160–177.

Pless, N. M., & Maak, T. (2004). Building an inclusive diversity culture: Principles, processes an practice. *Journal of Business Ethics, 54*(2), 129–147.

Pretty, J. (2003). Social capital and the collective management of resources. *Science, 302,* 1912–1914.

Price, R. H. (2006). The transformation of work in America: New health vulnerabilities for American workers. In E. E. Lawler III & J. O'Toole, (Eds.), *America at work: Choices and challenges* (pp. 23–35). New York: Palgrave Macmillan.

Rayner, C. (1998). The incidence of workplace bullying. *Journal of Community and Applied Social Psychology, 7,* 199–208.

Rubinstein, J. S., Meyer, D. E., & Evans, J. E. (2001). Executive control of cognitive processes in task switching. *Journal of Experimental Psychology: Human Perception and Performance, 27*(4), 763–797.

Schneider, B., & Bowen, D. E. (1993). The service organization: Human resources management is crucial. *Organizational Dynamics, 21*(4), 39–52.

Schön, D. A. (1987). *Educating the reflective practitioner: Toward a new design for teaching and learning in the professions.* California, US: Bass Publishers.

Svendsen, A. (1998). *The stakeholder strategy: Profiting from collaborative business relationships.* San Francisco: Berrett-Koehler Publishers.

Tarique, I., & Schuler, R. S. (2010). Global talent management: Literature review, integrative framework, and suggestions for further research. *Journal of World Business, 45*(2), 122–133.

Ulrich, D. (1997). HR of the future: Conclusions and observations. *Human Resource Management, 36*(1), 175–179.

Ulrich, D., Brockbank, W., Yeung, A. K., & Lake, D. G. (1995). Human resource competencies: An empirical assessment. *Human Resource Management, 34*(4), 473–495.

Vickers, M. R. (2005). Business Ethics and the HR role: Past, present, and future. *Human Resource Planning, 28*(1), 26–32.

Vohs, K. D., Mead, N. L., & Goode, M. R. (2008). Merely activating the concept of money changes personal and interpersonal behavior. *Current Directions in Psychological Science, 17*(3), 208–212.

Wade, R. (1994). *Village republics: Economic conditions for collective action in south India.* San Francisco: ICS Press.

Weaver, G. R., & Treviño, L. K. (2001). The role of human resources in ethics/compliance management: A fairness perspective. *Human Resource Management Review, 11*(1–2), 113–134.

Wilcox, T. (2006). Human resource development as an element of corporate social responsibility. *Asia Pacific Journal of Human Resources, 44*(2), 184–196.

Wilkinson, A., Hill, M., & Gollan, P. (2001). The sustainability debate. *International Journal of Operations & Production Management, 21*(12), 1492–1502.

Wimbush, J. C. (2006). Spotlight on human resource management. *Business Horizons, 48*(6), 433–436.

Wright, P., McMahan, G., & McWilliams, A. (1994). Human resources as a source of sustained competitive advantage. *International Journal of Human Resource Management, 5,* 299–324.

Wright, P. M., & Snell, S. A. (1991). Toward an integrative view of strategic human resource management. *Human Resource Management Review, 1*(3), 204–224

Yeung, N. (2010). Bottom-up influences on voluntary task switching: The elusive homunculus escapes. *Journal of Experimental Psychology. Learning, Memory, and Cognition, 36*(2), 348–362.

Zukin, S., & DiMaggio, P. (Eds.). (1990). *Structures of capital: The social organization of the economy.* Cambridge: Cambridge University Press.

Index

C. Machado and J. P. Davim (eds.), *Circular Economy and Engineering*, Management and Industrial Engineering,
https://doi.org/10.1007/978-3-030-43044-3

The manufacturer's authorised representative in the EU is Springer Nature Customer Service Centre GmbH, Europaplatz 3, 69115 Heidelberg, Germany. If you have any concerns regarding our products, please contact ProductSafety@springernature.com

Printed and bound by CPI Group (UK) Ltd, Croydon, CR0 4YY
30/09/2026
02225574-0002